241 O

# Christian ethics for a digital society /

241 O

6/19
TS

# CHRISTIAN ETHICS FOR A DIGITAL SOCIETY

# CHRISTIAN ETHICS FOR A DIGITAL SOCIETY

**Kate Ott**

ROWMAN & LITTLEFIELD
Lanham • Boulder • New York • London

Published by Rowman & Littlefield
An imprint of The Rowman & Littlefield Publishing Group, Inc.
4501 Forbes Boulevard, Suite 200, Lanham, Maryland 20706
https://rowman.com

6 Tinworth Street, London SE11 5AL, United Kingdom

British Library Cataloguing in Publication Information Available

Library of Congress Cataloging-in-Publication Data Available

ISBN 978-1-4422-6737-4 (cloth : alk. paper) | ISBN 978-1-4422-6738-1 (ebook)

∞ ™ The paper used in this publication meets the minimum requirements of American National Standard for Information Sciences Permanence of Paper for Printed Library Materials, ANSI/NISO Z39.48-1992.

Printed in the United States of America

# CONTENTS

# PREFACE

Digital technology directly impacts who we are and what we do; it ushers in abundant possibilities and indelibly alters how we form relationships and community. Current Christian approaches are often limited to an instrumentalization of technology that does not account for its significant social-shaping effect. Thus, we need an ethical language grounded in digital literacies and entwined with creativity and faith values. For Christians committed to a more just and inclusive society, an imaginative and serious ethical response to digital technologies requires looking beyond the immediate experience of software or hardware to how it was created, how our use of it shapes us, and what values are promoted across these processes.

How do we get inquisitive, ethically inquisitive, about digital technology? As a parent and educator, I have many more theological and ethical questions than answers when critically engaging digital technology. Most of my teaching and public speaking involves issues of sexuality and gender related to education, healthy relationship development, and violence prevention, specifically for teens. Digital technology, or even media studies and technological histories, was not an area of study I thought I would pursue. However, as digital technology became ubiquitous, I received more and more questions about the impact of digital devices on communication, social networking on relationships, information access on education, and so on. So I decided to turn my research abilities, time, and resources to ethical questions raised by digital technology.

This was a risky choice. I have never taken a course on digital technology. It is somewhat daunting to have to learn a new subject area with its own languages and history in order to ethically engage its impact. In this way, I have tried to "practice what I preach" in the text. I had to develop digital literacies as I researched and wrote. I am still learning. I guarantee imperfections will be present. A technologist may read this book and find that I am still a novice related to digital technology.

Literacy in any language, subject, or technology requires ongoing learning and cultivation of skills. Most digital literacy approaches are about teaching skills to use particular technologies so learners can become "producers" of digital content or better discern the quality of digital information—think making a website and knowing the difference between fake news versus substantiated journalism. In this text and for my own learning, I focused on a different aspect of digital literacy. The Christian ethic of digital literacy that I present builds on traditions of radical digital literacy and digital citizenship that argue for responsible use, reducing disparities in access, and attention to the impact of digital technology on relationship and community formation. The Christian part of the ethic grounds these concerns in historical and ongoing Christian visions of the common good and social justice for diverse communities and the earth.

Thus, I have sought to develop in this text a process that readers can use to hone their digital literacy as well as Christian ethical engagement. Starting from the point of everyday engagement, each chapter of this book unpacks digital technology issues that have significant ethical impact. I then bring each digital technological issue into creative conversation with theological understandings and scriptural narratives. Out of these conversations come ideas for large- and small-scale changes that readers can practice or research further as they engage a Christian ethic of digital literacy.

Unlike other Christian books about digital technology, the moral responses I suggest are creative and embodied rather than rule based. From the introduction onward, I promote ethics as play, creativity, and engagement rather than as sets of rules that are obsolete in a month or two given the rapid digital changes. In the United States, we increasingly hear about the need for personal practices of unplugging. Such suggestions are backed by research related to personal well-being and are helpful; however, that focus misses significant ways we are shaped as global digital citizens. As Christians in a digital society, we need a broader focus in order to consider the commercial, environmental, and cultural consequences of digital technology.

If we ask "What are the most pressing social issues related to digital technology use?" we must also clarify "most pressing" for whom? For some people across the world, gaining access to a digital device means the ability to have a business. For others, the need to unplug is urgent because digital social engagement is damaging their relationships and sense of self. Overall, the environment is being destroyed in some parts of the world because of mining for materials to make more devices, and elsewhere digital technology is saving the environment by reducing the need for travel and innovating smarter, energy-efficient machines. The most pressing social issue may in fact be developing an awareness of and response to these seemingly contradictory interconnections. First, we need to know what they are, and then we need to think more creatively and interdependently about how we can respond to them in local and global contexts.

Since this book does not provide quick fixes or formulaic ethical answers, I encourage readers to use their newly developing digital literacies to adopt an approach that technologists refer to as hacking. I do not mean hacking in the sense that we find it used in everyday media, which carries a morally negative connotation. Hacking as a practice can be enormously helpful in making systems stronger, raising public voices in a landscape dominated by corporations and governments, and adapting software and hardware to new purposes. The technical aspects of hacking may be skills held by the digital elite or underground, but everyday users also engage in hacking. We creatively adapt and modify technologies to fit our community needs. I suggest that hacking includes the ethical call to gain access to the ecosystem of digital technologies and define the vulnerabilities to be patched as the perpetuation of social inequalities and injustices. Think of it more like pounding swords into plowshares!

Similar to digital initiatives such as crowdsourcing and open source software, this book is a cooperative creation. Though I wrote the final product, I hope someday to write a book as an open source project that includes all the conversations leading up to and through the writing process and all those stemming from it captured in one multipurpose platform. Until then, I have this analog opportunity to reference the scholars, religious leaders, and students who have made this book into a richer resource than I could have ever developed on my own. A number of colleagues have served as my conversation partners related to different aspects of the text. Some are named in endnotes because their own scholarship directly impacted the text. For friends and colleagues such as Melanie L. Harris, Elias Ortega-Aponte, and Shawn Spaventa, I am grateful for the many times they listened to my ideas, challenged my approach, and supported this project. Along the way, other colleagues and editors provided opportunities for early forays into the subject matter, which gave me the courage to write a full book. I thank David Maxwell, Betsy Shirley, and Sarah Stanton. And though our relationship is new, Rolf Janke, my editor at Rowman & Littlefield, provided the last push of support and encouragement to bring this book to publication.

I am privileged to have a wonderful foundation in theological education, which made it possible to engage a completely new subject area and re-envision my own disciplinary approach. Of course, this meant I had to become a learner again in the truest sense of the word. There may be places where readers feel overwhelmed by the technological language, but I try to provide concrete examples and unpack the complexities when possible. I could not have honed the book approach, topic, or scope without significant assistance. In spring 2017, I taught a seminar titled Ethics of Digital Technology with ten students, ranging from PhD to undergraduate. I planned the first half of the syllabus, and the students planned the second half. Together, we did a deep dive into digital technology, creation, and ethics. The turn to an ethic of digital literacy is taken from the evaluation of that class when one student, Lisa Asedillo Pratt, said, "This was like a language immersion course." Many of our class conversations reshaped my own approach as students introduced me to new resources, asked questions from their community location, and wrestled with personal and structural ethical issues. For the generous support of this course and my ongoing research, I thank the Everyday Ethics initiative at Drew University, funded by Edward and Barbara Zinbarg.

The second major influence on the project is my colleague and developmental editor, Darryl Stephens. This book has an organizational clarity and coherence that would not have been possible without his feedback. He has helped me become a more patient and accurate writer while also challenging me to push my ideas into new ethical territory.

Both these aspects make this book a better experience for the reader and hopefully a stronger contribution to the field of Christian social ethics.

In addition to direct assistance, I have had colleagues in churches and social movements across North America who have talked with me about this book and even read portions of the drafts. I am particularly grateful to Christian educators and pastors who reached out to me during the writing of this text to ask for resources and engage in conversation. I owe a particular note of gratitude to Shannon Clarkson and members of RSAC (ad-hoc group on ending racism, sexism, ageism, and classism), who invited me to facilitate a conversation on the book. This group piloted discussions related to the excurses found between each chapter, adding new questions and providing valuable feedback.

During the editing of the book, I unexpectedly joined a consulting team, with a longtime colleague and digital collaborator, Kristen Leslie, to teach Navy chaplains about issues of young adults, religiosity, and digital technology. I was welcomed into a new context of ministry and surprised by the need for ethical leadership related to digital technology. That experience took me across the globe and reinforced the desire to address the deep changes that digital technology has on our relationships, theologies, and ministry.

While digital technologies shift our lives to fast-paced forms of communication and tantalize us with ever-present distractions, the labor of writing a book is a slow and tedious process. The labor is made significantly less arduous by the advent of digital technology, including cloud computing and access to information. Unfortunately, there is still a significant amount of solitary time spent reading, writing, and editing. For the gift of that time, I thank my family—Brian, Eva, and Isaac. Without their support, which has been both mundane and monumental, I would not have written this book.

# INTRODUCTION

**D**igital technology and social media can seem almost magical when we step back and consider how rapidly the technology develops and integrates with our daily lives. We can see our whole family on a video call from halfway around the world. Depending on the settings, our phones can remind us where we need to be and even provide directions to the location without our even asking for the information. Smart assistants, such as Siri, Alexa, and Ok Google, use artificial intelligence to do our shopping, answer our searches, and provide entertainment all at the command of a voice. The rapid change can be exciting and provoke imaginative possibility. For some, anything that makes life more efficient and productive is considered good. For others, the pace and integration are cause for panic—moral panic—as relationships to each other and our world are redrawn. In *Overcomplicated: Technology at the Limits of Comprehension*, Samuel Arbesman argues that this leaves us with two responses: "crippling fear" or "worshipful awe." He says, "Both prevent us from confronting our technological systems as they actually are. When we do not take their true measure, we run the risk of losing control of these systems, enduring unexpected and sometimes even devastating outcomes."[1] To move beyond fear and bring some realism to our awe, critical engagement with digital technology is desperately needed as a daily ethical practice.

Too many of us are technologically dependent and not technologically savvy. We have much to learn about how digital technologies work so that we can more responsibly interact with them. We are not well informed, and thus we are at risk for being morally malformed by digital technology. To assess digital technologies, we need a healthy level of skepticism toward technologies as well as an acknowledgment of their benefits. We can be neither wholly positivist, those who view any form of technology as moving humanity forward, nor negativist, those who believe technology stifles humanity's growth. Either approach can be correct, but neither is always true.

Our understanding of digital technology directly influences our ethical response to it. When we instrumentalize digital technology in ways that narrowly conceive of our relationship with it, the ethics that arise in response seek control of technology, such as individuals restricting personal use. Many instrumental responses are obsolete as soon as we proclaim them. On the other hand, when we ascribe infinite power to digital technologies, leaving little room for human action, the ethical response can be reactive and hopeless. When we position digital technology as the antithesis of humanity (it/IT vs. us), we deny the human, relational impact and goods that come from technological development. The ecosystem of digital technology requires that we cultivate relational and imaginative approaches rather than moral absolutes. It requires an evolving sense of both ethics and technology. Digital technology is responsive, adaptive, and networked. Traditional views of ethics grounded in absolutes or calculations will not serve us well in a digital world. Rather, we need ethical approaches that embrace growth, interdependence, and creativity. These ethical approaches mean we will need to lean into specific aspects of scripture and Christian theological traditions that have not been as prominent in traditional Christian ethics discourse.

Digital technology is rapidly changing our lives and presenting new ethical questions. Yet many of us engage digital technologies without considering how their design is reshaping the ways we might think about our relationship with God, Christian practices, care for creation, or work toward a more just society. When we think theologically about technology, we consider how values and practices are promoted or limited by digital engagement. As a Christian ethicist, I focus on Christian values and practices in this book, but my hope is that the approach raises similar questions for other religious traditions as well as philosophical or humanistic value concerns. Instead of simply worrying about the next technological gadget, digital app, or platform, I shift the focus to what Christianity has to offer a digital world. Living as a Christian in a digital society calls us to rethink faith values and practices. The purpose of this book is to begin a conversation that considers the massive ecosystem change that digital technologies push in our lives through a focus on the ethics of everyday living.

## CHRISTIAN ETHICS AS CREATIVE MORAL RESPONSE

Ethics is about asking questions and seeking moral responses. Christian social ethics unpacks how social and institutional structures, in this case digital technologies, promote or prevent a more just social order for those who experience oppression based on race, gender, economics, the environment, and so on. When it comes to digital technologies, Christians often ask whether technologies can be used to bring more people to God or how we see God in our technology. I ask, *what does God require of each of us to be and act in a way that promotes Christian values* in all that we do, including the digital technologies we develop and use?

To answer this question, we need to shift ethics from a highly rationalized, independent calculation to ethics as a form of social literacy, developing and adapting throughout our lives. When it comes to technology, we more often than not devise "rules of use." Digital technologies, however, require an engaged learning approach that matches their fluidity and adaptability. That concept will have ramifications for how we understand ethics. Ethics is often considered the domain of adult decision making and the implementation of preexiting value sets. In actuality, ethics is something we all do, all the time. Ethics is extremely contextual even when we can see certain values exhibited across experiences, cultures, or time periods. Ethics is a process of growth, not a static set of rules. Ethical growth is an ongoing process for which we develop capabilities throughout life.

Ethics focuses on responding to two questions that intersect and affect each other: *who ought we to be* and *what ought we to do*? In Christian ethics, we ask these questions in relationship to our understanding of God and the history of the Christian community. Christian ethics usually addresses these two questions by applying principles, defining and promoting character development, identifying duties, or calculating ends and consequences. Each of these approaches assumes an adult is the "ethical actor" or that adults are enforcing "ethical behavior." Historically, in Western Christian ethics, Anglo, religious, educated men in particular were viewed as the most morally capable individuals. Theologians claimed they demonstrated the highest forms of rationality, independence or autonomy, and experience. We still use these defaults when considering who counts as a moral decision maker. We assume one needs to know how to explain one's choice in a logical and informed manner for the decision to "count" as ethical thinking. We assume independence is needed to exercise such decision making and that previous experience better equips us for difficult moral situations. However, none of us make moral decisions devoid of emotion, outside the context of relationships, or directly informed by a store of ethical considerations. Each time we face a moral decision or are called upon for ethical response, we do so with all of our limitations and potentialities, including being embodied, thoughtful, emotional, relational, and experimental.[2]

We see a much more dynamic form of ethics when we change the primary subject in ethics from an adult to a child; this shift provides a more fluid and adaptable view of ethics that matches better with digital technology. Children have been considered premoral, or in the process of becoming moral people, which suggests, then, that at some magical, future age they will eventually "know" what decisions they are making and we can hold them accountable for those decisions. A number of theologians and ethicists are crafting a new understanding of ethics by looking toward children.[3] They have found that children evidence moral decision making on a daily basis and in quite complex ways. Because children are sometimes incapable of explaining or analyzing their actions, adults see them as happenstance or mimicry.

For example, is it a moral action if a three-year-old at the snack table shares her snack when she notices another child is sad because she does not have a snack? The three-year-old cannot explain concepts like equity and justice in response to disparity. She may be able to say, "I don't want my friend to be sad, so I shared my snack." She may even devise a way to do this without the teachers knowing if the school does not allow sharing of food, which is an even more complex decision because she risks being punished for what she believes is helpful to another student. The rest of the students who do not share are also involved in a moral response as are the teachers and the preschool based on the rules it has established. Ethical response requires a good deal of cognitive assessment of a situation, including emotional and social evaluation of circumstances and responsiveness to others we encounter. Children engage in these processes even when they are very little.

Like any kind of development, ethics requires growth. Ethical growth happens throughout our entire lives. It is not simply a set of rules or an approach we learn as children and youth that then makes us "moral people" as adults. We struggle with ethical issues throughout our lives and are constantly searching for ways to respond. John Wall, based on his investigation of ethics related to children, describes ethical response in this way: "It creates received historical and social meanings into new worlds of meaning over time and in response to others. It deals in moral tension and disruption as selves confront their own narrative diversity and the otherness of others."[4] Our everyday lives are filled with moral decisions. Living is an ethical response because we are constantly confronted with "others" with whom we are in relationship. We often fall into patterns of ethical behavior so we think less about each interaction, but each action is nonetheless shaping ethical growth. As we meet the "other," we meet God. These ethical encounters are the basis of a cocreational relationship with God that embraces the diversity of our experiences and calls forth various forms of knowledge, including embodiment and interdependence with creation.

The newness of digital technologies and the shifts it ushers into our social patterns often disrupt historical and social meanings, ways of acting and thinking toward others, that adults have stopped noticing. Wall does not investigate digital technology. However, he does turn our attention to what is lost when we rely too heavily on ethical formulas. Wall invites us to unpack these patterns and interrogate how we encounter the other in mundane everyday situations, especially for children. He is concerned that "fixed principles, laws, and virtues have ever since dominated over children in particular and over imagination, interdependence, and change in general."[5] For Wall, "Ethical thinking is an art."[6] It requires a sensibility to more than a set of rules. That is not to say rules or moral principles are not helpful, but if they foreclose on the decision or one is required to exclusively assess them in a logical–rational manner, then we exempt individuals from their own moral process. Moral growth is about responding creatively to otherness.[7] We expand our "moral horizons" every time we have to imagine what is different about another person, animal, plant, or the earth and what is needed in response to that difference.[8] This description of moral decision making is artistic. It is about finding better solutions, not one right answer. It also promotes a reciprocal relation between the process of response and reception or reaction rather than a linear finality.

A focus on the everyday use of digital technologies might seem counter to the large-scale changes digital technology has triggered. Emilie Townes in the concluding paragraphs of her book *Womanist Ethics and the Cultural Production of Evil* reminds us that "this is not a quest for perfection, but for what we call in Christian ethics the everydayness of moral acts . . . the everydayness of blending head and heart; the everydayness of getting up and trying one more time to get our living right."[9] Relating ethical issues to everyday living does not mean we ignore the magnitude of the problems we face; rather, it requires a knowledge of how those problems relate to the everyday so that we can be intentional and transformative in our responses. As we wrestle with the shifting moral landscape of digital technologies, I assume a bias toward action, an explicitness about everyday engagement. Accepting that "ethical thinking is thus in its most fundamental sense creative or poetic" does not mean we avoid concrete actions.[10] It does, however, suggest more fluidity and responsiveness in an ethical approach. In response to digital technologies, I explore the need to increase diversity, engage cocreative responsibility, cultivate attunement, deepen our practice of metanoia, and hack our social worlds in a participatory manner to reduce inequalities. Each chapter will address one of these issues by engaging theological and scriptural traditions, considering design and use of an aspect of digital technology, and proposing ethical responses.

These ethical approaches are based in an ongoing process of moral growth that is meant to expand moral response to the otherness that we experience in the use, creation, and extension of digital technologies. Each of the approaches results in a variety of concrete practices that are worked out in particular situations based on the contextual location of the individual.

Underlying the moral approaches discussed throughout this book is the theological awareness of God with us, calling us to moral response. In writing about ministry, Richard M. Gula notes, "Incarnational theology affirms that 'earth is crammed with heaven' so that every experience, if given a chance, can speak to us of God."[11] As Christians, our digital experiences and the ecosystem of digital infrastructure speak to us of God. In some cases, they evidence our lack of response to God, and in others they evidence new ways to respond. The rapid change of digital technologies has in some cases swept us up into patterns of thinking and acting without moral reflection. Other times, confusion, lack of awareness, or even blissful ignorance of the complexities of digital technologies means we do not have the resources to adequately reflect on moral implications of these technologies.

We must search for those who are engaged in creative moral responses. I provide details about a few examples, such as the Algorithmic Justice League, e-waste artists, and digital humanities educators. The presentation of these responses helps us answer, *How on an everyday basis do we begin to notice our moral responses and cultivate new meanings out of these encounters?* Creative moral response is nurtured by a socio-moral pedagogy, what liberation theology terms praxis, defined as "reflection and action directed at the structures to be transformed."[12] In the tradition of liberation theology, critical awareness of one's material condition caused by oppression related to economic, racial, or gender disparity is key to transforming the social structures that perpetuate oppression. With regard to digital technologies, I note the ways in which we may experience oppression as well as participate in it on individual, communal, and systemic levels.

Critical awareness implies learning how systems work, developing a socially minded curiosity that often starts with observations about personal experience and connects to larger social patterns. This process, termed conscientization, is relational and thus cultivated in community. For example, we reflect on social network behaviors, ours and others' in our network, considering how the behaviors are shaped by and respond to systemic forces such as network design, profit motivation, use of personal information, and so on. Knowing how these built-in structures restrict or promote expressions of self and relationship building with others enables us to more actively engage moral responses that may cultivate attunement as embodied digital selves. By engaging in this type of praxis, we increase our digital literacy and capacity to morally respond to new aspects of digital technological change.

## DIGITAL LITERACIES AS PRAXIS

To live out the moral approaches to digital technologies suggested in each chapter—expanded diversity, cultivated attunement, responsible cocreation, transformative metanoia, and ethical hacking—we need to enhance our digital literacy or literacies. Literacy education is often the domain of educational systems from preschool through high school. Similar to literacy related to analog print and text, K–12 educational systems are now introducing digital literacy education. We associate literacy with the ability to read and write a language system. That is literacy in its most basic sense. To read and write, we also need a variety of other capabilities, such as identifying social context, being aware of relationships and audiences, and deciphering differences among types of text or conversations. For example, reading a poem requires different capacities than reading a newspaper article, as does writing a sermon in comparison to writing a thank-you note. Similarly, then, digital literacies must also account for the variety of "practices of communicating, relating, thinking and 'being' associated with digital media."[13]

Unlike analog literacy, what we traditionally think of as reading and writing, or language arts curriculum, digital literacies are overwhelmingly learned on a peer-to-peer and informal basis.[14] This is partly due to an early bias against use of digital technologies in classroom spaces and a slower adoption by adults of digital technologies, who often want to be the expert in the room before they use a medium or teach a subject. These practices are shifting as educational systems have realized we should not leave children to their own devices (literally). Digital literacies education is needed so that we can "even out children's and young people's access to media texts, technologies and cultures. It [education] can develop an ability to reflect on ethical issues, on meaning, on power."[15]

In a digital age, we *all* need to develop digital literacies. These movements need not be contained to formal educational systems engaging only children and youth. In fact, that is the primary motivator of this book. *A critical part of moral praxis in a digital age is cultivating digitally literate communities.* When we invest in "developing creative imaginative talents there are individual and social benefits."[16] Digital technology is a mediator for human interaction similar to language. Some might say digital creates a new language, but it is more than a communication system built on symbolic code because it involves use of devices.

All technologies, whether digital or not, mediate different kinds of *actions, meanings, social relationships, ways of thinking*, and *social identities*. In *Understanding Digital Literacies*, literary and media theorists Rodney Jones and Christoph Hafner explain each aspect. *Doing* relates to what technology allows us to do in the physical world that we could not do without it. For example, a hammer allows us to drive a nail into wood that we cannot accomplish with a bare hand. Technology allows us to make different kinds of *meaning*. Telling a story in person is different than writing about it in a newspaper or making a television show about it. Technology allows us to create different kinds of *relationships* by changing forms of interaction. This book allows me as an author to "talk" to a larger audience in a long-form engagement, which I could not do in person or even using video. The way we experience and *think* about reality is changed by tools. The mediation of technology often affects more than one aspect, such that changes in doing can impact thinking. Unlike a hammer, which changes how we hit a nail, a mobile phone allows instantaneous communication with a person in a different geographic location and time, traversing time and space to an extent that shifts our notion of reality. Finally, technology affects who we are, or *being*, related to how we organize our social identities and awareness of audience.[17]

Literacy as it relates to various cultural technologies, from tools such as a hammer to any world language to the expanding digital ecosystem, requires attention to all five aspects of how we are mediated by technologies. That is to say, literacy refers to more than how to use a particular language or machine; it is "the ability to creatively engage in particular *social practices*, to assume appropriate *social identities*, and to form or maintain various *social relationships*."[18] Based on the diversity of these needs, I am following other media theorists and using the plural of the term—literacies. Digital literacies include knowledge and experience with concrete machines; information; software operation; and written, audio, and visual texts. These include but are not limited to literacies related to photovisual, reproduction, branching, informational, socioemotional, and processing capacities.[19] Digital literacies encompass all of these aspects. For example, in chapter 1, we have to explore how algorithms are created and how they function and impact the use of digital technology in order to respond to algorithmic bias. Because algorithms are central to the operation of most digital technologies, there are multiple aspects of digital literacies that are needed to understand them.

With respect to moral formation and moral response, I argue that we need to go a step further. We must develop digital literacies with an ethical intention in mind. The point is not only understanding but also praxis, in the sense of action and reflection for transformation. Jones and Hafner refer to this formation of digital literacies as critical digital literacies. They state: "We cannot avoid using various mediational means to take action in the world. However, the more we know about how these media work, the better we can become at hacking them through selective appropriation, adaptation, modification and mixing to fit our own purposes and promote our own agendas."[20] Throughout the chapters, I invite the reader to alter the landscape of digital technologies toward moral formation that promotes diversity, responsible co-creation, attunement, metanoia, and equity. The shift is more than critical; it is transformative and liberative.

## MORAL APPROACHES AND DIGITAL TECHNOLOGIES

The book follows an ethical shift, seeking liberative elements found in the everyday both in its content and form. The arc of the chapters sketches various components in their particularity that comprise a Christian social ethic of transformative digital literacy. Each chapter raises an ethical issue that takes seriously the various ways digital technologies contribute to as well as combat growing inequalities related to gender, race, economics, and the environment. I begin each chapter by engaging theological and moral concepts implicated by a central ethical question related to a specific aspect of digital technology. I then explain current technological, media, and social science research on development and use of the technological aspect. After a richer understanding of the technology is explored, I connect the technological issues with ethical responses based on theological and moral resources presented in the first section. Each chapter concludes with specific ideas for concrete change. Throughout the book, I embed excursus—similar to a website pop-up—as an extension of the main text to provide a spiritual reflection on the ways we navigate embodied, digital moral life as part of the Christian community.

Chapter 1 locates us at the heart of digital technological change and ethics. Algorithms are the sets of rules by which computer programs complete tasks. Algorithms manage data production, collection, and interpretation—learning from behaviors and predicting outcomes. For example, when we shop online for a new book and then log into a social network and see that same book advertised on the page, we might find the connection a bit too coincidental. It is not coincidental at all. Algorithms use more data points than humans can calculate to fulfill their designed goals. Research related to the effect of algorithms shows that they perpetuate bias along age, race, gender, sexuality, and ability. One way they do this is by creating increasingly personalized profiles for users that shrink the wide expanse of the internet in the name of efficiency, accuracy, and profit. If we compare the diversity of what the internet offers and what real users experience, we find that most of us live in very closed online spaces regarding religion, news, geographic location, and so on. In chapter 1, I argue that the closing off of our worlds is counter to the diversity of God's creation and our role as cocreators. A closer look at the biblical story of Babel situates a theological understanding of difference as part of God's creation and how it relates to language, even the language of algorithms.

Chapter 2 zooms in on the user and shifting relationships with the self, others, and God through interactions with digital technologies, especially on social networks. Some claim we are replacing God with techno-demigods. It is true that many digital technologies shape our desires, often via advertising toward consumer goals. Research confirms that digital technologies rewire our attention and promote a kind of multifocal and networked way of knowing the world. A networked and collaborative approach to "knowing and being" may promote inclusivity and provide innovative ways to *be* Christian that are supported by relational, trinitarian understandings of God. I explore attunement as part of moral formation toward relationality and connection rather than isolation and distraction.

The turn to the user situated within social networks is then expanded in chapter 3 to consider what becomes of the data we generate and how we are shaped by forces regulating and interpreting our data. Chapter 3 raises critical ethical questions about how digital technologies archive our every post, picture, email, and so on. Whether it is personal, a loved one or a complete stranger, our actions, images, and secrets are increasingly documented, shared, and stored. Some wonder if we will ever "die" given the potential for immortal digital lives; some argue the internet should be reprogrammed to forget. In the Christian tradition, forgetting may impede forgiveness, which is a key aspect of lived faith that positively enables the human capacity to make change. In fact, forgiveness is a necessity for a sinful people seeking eternal life with God. When we zoom out from individual users, we begin to see an ecosystem of digital surveillance that perpetuates larger social problems deforming our moral communities. Thus, forgiveness is one step that can lead to metanoia, a transformative shift that liberates us from the individual and systemic sins of today's world. Similar to a reshaping of ethical orientation around attunement, this chapter considers the need for metanoia in our response to digital technology.

In chapter 4, I explore how archiving issues are connected to material realities of the environment. The use and consumption of digital technology requires significant costs from energy sources and environmental resources. This chapter addresses concerns of e-production and e-waste. As the number of devices and need for high-speed internet grows, so do server farms and outdated electronics. Reduce, reuse, recycle is often a motto associated with plastic drinking bottles, not mobile phones, tablets, or computers. Technological advance in a human versus nature model can be theologically rationalized as human dominion over creation. Shifting to theological models of cocreators requires ethical responsibility to develop ecologically friendly digital technology as well as alter consumer practices to promote biodiversity. Practices that develop aesthetic and embodied connection to the earth and digital technologies enhance humanity's role as cocreator and contribute to the diversity of creation.

The concluding chapter focuses on the cultivation of digital literacies that move users from a position of consumption or reception to one of creation. This chapter highlights how participatory communities' moral response to digital technologies is a form of "hacking" with the potential for ethically transformative acts like pounding swords into plowshares. This final conversation brings together the theological explorations of hacking with key components of digital literacy and moral growth as creative actions and orientations. The various moral approaches in chapters 1 through 4 weave a foundation for hacking approaches in chapter 5.

The excursus placed at the end of each chapter provides a reflective pause. The excursus also reshape the contour of a linear textual argument by inserting and disrupting the analog pattern of a book. When reading in a digital environment, we can now click on a word and be linked to a new set of stories or highlight a word and see its definition and function throughout the text. Each excursus begins with a scripture passage that frames a more personal reflection on the chapter theme and connects it with a specific social issue. These, like a link off a web page, could be read as stand-alone pieces or completely skipped depending on a reader's interest. Read along with the chapters, they are an example of the everyday ethics approach and the action–reflection method discussed above. Their presence in the text call the reader to make a shift, one that finds similitude with digital spaces.

## "WE" AND MORAL PROPORTION

Digital technology access, use, and production result in global disparities. And yet more people than ever have access to smartphones and the connections they provide. The digital divide has been shrinking though education level, household income, age, disability, and geographic location still contribute to differences in use and access to digital technologies.[21] Those divides are exacerbated when we compare developed countries with developing countries. Within any given country, there are also divides. Addressed throughout the book are ways that digital technologies contribute to inequalities. On the other hand, equally acknowledged will be the ways digital technologies are used to reduce inequality and oppression. The difference does not simply lie in "how we use" the technologies; it includes how they are manufactured, designed, and powered.

When I write "we," I am invoking a potential and yet general demographic readership of this book. I am writing for those who have the means to be wired in a digital world. This description includes people of a variety of racial and ethnic backgrounds and gender identities, primarily in the United States, though not entirely. It assumes a socioeconomic status that is not dependent on a particular income level but carries educational and global geographic privilege. There is a privileged point of view the reader *and* I bring to the issues presented here.

Who, then, bears moral responsibility and to what extent? Digital technology affects everyone across the globe, even those who choose not to or cannot afford to be users of digital technology. That said, each of us bears a different responsibility in moral proportion to our privilege. For example, on a travel seminar in El Salvador, I spoke with a woman vendor who uses a mobile phone as her primary form of communication. She uses it to know if her vendor space is available, to keep connected with family, to purchase more goods, and so on. Mostly, she relies on locations that provide free wireless access to cut down on charges for data usage. She pays between 15 percent and 25 percent of her monthly income for this digital connection. I, on the other hand, have access to wireless in my home, my workplace, and most public locations and unlimited data for less than 2 percent of my monthly income. Our monthly incomes are vastly different, and so is my access and usage of my digital connection. I bear a greater responsibility for the moral effects of digital technologies because I have a larger share of the privilege. Economics is not the only way to analyze moral proportion. Throughout the text, I will present a variety of differences that contribute to digital disparity that thus necessitate more adept and pliable moral approaches.

The "we" includes me and the majority of readers, though not all of the time, and thus engaging the text with one's own social location in mind is the ethical task of the reader. Attention to the self as a reader and myself as an author is part of cultivating literacies. In this case, the practice is analog in relationship to this text. Developing literacy means we take seriously how technology shapes culture and community and vice versa, whether that is reading a book or designing a website.

Recent writings related to technology and ethics range from books focused on social criticism to how-to guides for Christian ministries. The majority of authors in this category are white, male evangelicals who share a theological approach with one another that places God's revelation and intervention as central to their investigation.[22] Other authors delve into philosophical histories related to technology and mine those traditions for current direction and theological critique.[23] The Christian social ethics approach of this text keeps central the role all Christians have to respond to inequalities evidenced in everyday lives held in place by less visible social infrastructure.

## MORAL MEANS

We often hear about the negative effects of digital technology. At the same time, digital technology has become rapidly integrated into our everyday functions in positive ways, and we often uncritically, even enthusiastically, adopt these innovations. In many cases, Christian communities respond to technological advances in an either-or manner—rejecting technology as a distraction from God's plan or embracing technology as a new avenue to do God's will. Somewhere between moral panic and blind faith lies creative and critical ethical engagement. For Christians committed to a more just and inclusive world, imaginative and serious ethical response to digital technology requires looking beyond the immediate experience of software or hardware to how it was created, how our use shapes us, and what values are promoted. Pulling the curtain back to "reveal the wizard" demonstrates how digital technology can perpetuate human divides and ecological degradation. Yet equally possible is a story of reengineering community for a more just world.

A long-form book may seem like a strange way to engage the complexities of a fast-changing digital reality. From the first to the last chapter of this book, I invite the reader to consider the radical and abiding impact digital technology has on our lives. There are times when 140 characters are sufficient and other times when fifty thousand words are needed to explore ethical issues and digital technology. To consider a fuller view of the ecosystem of digital technology, we need sustained, long-form conversation about how Christians are being shaped and changed and consideration of what resources Christianity has to offer in our current techno-culture. The conversation may happen on social media in short form as well as in a full-length book. In most cases, both, and then some, are required.

# 1

# PROGRAMMING FOR DIFFERENCE

From ten thousand feet, the digital world may look like a grand bazaar, but when we zoom in, our technologies have been so personalized we rarely see this diversity. In other words, as our lives become more individualized and the possibility of information becomes increasingly globalized, we rarely question what we do not see. In fact, even if we tried to interpret what creates these circumstances, it can be nearly impossible. Algorithms operate beyond most human users' awareness. Furthermore, the mass amount of data created by digital technologies is too much for humans to process. Because algorithms are used for this task, they mediate how we interact with the world. Algorithms appear to be neutral, logical processes; however, the design, implementation, and outcomes often result in disparate impact[1] or bias on a collective level. On a more individual level, algorithms govern data and generate the information we see, limiting our options in helpful ways but also with unseen negative consequences.

For example, consider the use of online map services, which rely on algorithms. We have become accustomed to typing in a general location, relying on the map service to specifically locate it, calculate traffic and construction (in real time if we use a GPS service while driving), and provide the shortest route. Gone are the days of getting lost and stopping at a gas station and unfolding a paper map from the glove box to find the way or using a AAA TripTik made out of multiple map sections, creating a spiral-bound trajectory of one's journey, noting government-reported road construction—with no real-time updates, accident reports, or alternative routes. GPS data and algorithmic map systems are definitely an improvement if one values certainty and efficiency over exploration and chance. Most of us do not expand the map layout to consider other routes or learn about landmarks and natural resources surrounding the selected route. The allure of the moving arrow draws the user's attention to move from point A to point B without hesitation. These behavioral changes, neither purely good nor bad, demonstrate a radical shift in the everyday behavioral feedback loops upon which we base our self-understanding. In this digital context, intuition, perception, and imagination are increasingly replaced by measurement.

The wide reach of digital technology beyond past boundaries of time and space promises an expansion of difference in the world with more information access, global connections, and capacity for larger networks. All technologies mediate interaction between self and other or self and the world. In this sense, technologies fundamentally change the kinds of societies we can build. The light bulb, for example, changed sleep patterns, work hours, and community safety. It altered social structures. In other words, technologies are not concrete material objects we completely control. They shape us as much as we shape them.

Algorithmic impact both in personal use and related to social programs will be addressed in this chapter. Digital personalization is the result of algorithms used to process and interpret the massive amounts of data generated by digital technologies. When we log on, we are more and more dependent on software platforms that utilize algorithms that sort through data to make it manageable. Eli Pariser has coined the term "filter bubble" to describe this phenomenon: "Together, these engines create a unique universe of information for each of us . . . which fundamentally alters the way we encounter ideas and information."[2] Similar to our use of language, algorithms constitute an abstract code or system that creates meaning and makes data intelligible, sifting through data points for connections. These connections turn into information and process outputs that humans now interpret as knowledge, which is information transformed to solve a task or combined to create something new.[3] There are multiple algorithms as there are multiple languages. They are increasingly used for the purposes of personalization and to perform tasks that limit human creativity and knowledge growth.[4]

Data production, collection, and interpretation are key aspects of digital life and integral to an understanding of algorithms. The middle section of this chapter will focus on these issues and specifically address the limiting of difference related to predictive analytics. When I email friends and then log into a social network and see their most recent posts at the top of my feed, I might marvel at the coincidence or wonder who is watching me from behind the black mirror of my mobile screen. This experience raises questions about how algorithms work. What produced data do they collect and what rules do they follow when interpreting that data? What values are promoted as algorithms shape the information we are given by our devices and social networks? Assumptions abound about the objectivity of computational processes. However, algorithms are built by humans and carry human intentions, sometimes conscious and other times unintentional. Not only must we consider questions about the meaning of information and how it shapes our environment and self-perception; we must also explore how the information to which we are exposed is being limited by algorithms. Do algorithms lead to greater understanding across differences?

In other words, we are told the internet is a gateway to radical difference while algorithms work unseen like the architects of Babel to narrow languages and create a personalized tower. Algorithms are meant to be hidden and often work in ways even the original code writers cannot foresee. They produce outputs that rely on predictions called predictive analytics. Users tend to respond to predictions as fact. Algorithms are the key to making sense of the information generated by the Internet of Things—computing devices built into everyday devices that connect to the internet, allowing them to send and receive data. Mobile phones and all the applications they hold are the most obvious and the most utilized digital devices in the United States. Now, we also have coffee machines, thermostats, house-lighting systems, alarm clocks, televisions, refrigerators, cars, watches, fitness bands, and so on with built-in internet connections. We are increasingly surrounded by devices that use predictive analytics based on algorithms to guide our behavior. Predictive analytics also affect human policy decisions and practices when outsourced to algorithms.

Diversity, or existence of differences, is a defining feature of creation. The Scripture story of Babel is an example of God affirming difference, in this case evidenced by language. Likewise, Pentecost, where the Holy Spirit brings the gift of understanding in one's own language, offers another example of affirming differences.[5] Part of created difference related to humans is the diversity not only of language but also of the very differences exemplified in humanity, including race, gender, personality, and physical and mental ability. We must develop ways to continue to expand and understand this valued aspect of creation.

Exploring the Tower of Babel narrative provides a grounding scriptural and theological resource for this chapter. Various scholars utilize the Tower of Babel to discuss a variety of theological approaches to dealing with the mass amounts of information and communication changes brought about by digital technology. I explore their work as well as consider how the affirmation of difference at Babel provides guidance when considering the necessity of promoting diversity in digital networks. One of the primary claims of this book is that everyday ethical living in a digital world requires curiosity and basic literacy with how digital technology functions. Awareness and understanding of algorithms is a necessary component of digital literacy, including how data is used, information is created, and predictive analytics are promoted. Learning these aspects of digital technology help us identify when they promote difference and when they erase it. From users to policy makers and programmers to tech companies, we all have a role to play in disrupting algorithmic bias.

## BABEL: VALUING DIVERSITY

Many Christian and some secular writers who engage issues of technology have turned to the story of Babel.[6] For them, it represents a biblical injunction against technological overindulgence in information and uninterrogated cultural shifts. Many share a reliance on an interpretation of God's action at Babel as a kind of punishment rather than encouragement. For many critics, Babel represents the fallout when humans try to use technology (building) to wrangle the cacophony of information and people into a singular connection seeking to become God or godlike. Thus, many writers conclude we need to turn away from technology. That interpretation misses a key feature of the story. As Míguez-Bonino suggests, Babel[7] is better understood as a story about God's gift of difference and Pentecost,[8] a story about a movement toward understanding this difference.

Algorithms are a bit like the building of the Tower of Babel (Gen 11:1–9). The Babel story addresses use of technology to build and the symbolism of language. At its core, the Tower of Babel is about difference and God's response to human efforts to erase difference. Those seeking to build the tower at Babel forced their language on the various cultures and people present to achieve their goal of building the tower. In destroying the tower and scattering the people, God denied a false unity based on sameness (one language and one political power) and gave people the freedom of multiple languages, geography, and political power. Thus, it is not the technology of building that God condemns but the use of a communication tool like language to erase difference. If Babel is about the affirmation of difference, then we are called to celebrate diversity, like that of language, viewpoints, and geographic location. As we develop, shape, and reshape technologies, are we doing so in a way that encourages difference and maximizes diversity similar to Babel? Or are we reducing diversity?

Christian ethics scholarly approaches to digital technology are few in comparison to other applied subject areas, such as war and peace, economics, climate change, bioethics, racial justice, or gender equity. Interestingly, the majority of these works mention the story of Babel. The first and most formal approach can be found in Brian Brock's *Christian Ethics in a Technological Age*.[9] Brock in his chapter "Seeking Christ's Concrete Claim" reinforces the presence of various technologies throughout history as well as the shift to modern technology as a dominant form of power and meaning making. He suggests the model of rationality leveraged by modern technology distracts users, leading them to seek fulfillment and salvation via technology. He wants to disrupt technology's hold, "not to renounce all modes and forms of technological rationality but to desacralize it."[10]

It is within this project that Brock discusses Babel. Brock juxtaposes the technology or building of the Ark with that of the Tower of Babel. He concludes that humanity can use technology for good or evil. The determining factor in such an evaluation is based on how the technology serves "God's work of redeeming creation from sin."[11] Babel, Brock concludes, is a misuse of technology, a human-initiated effort to overcome God. For him, use and creation of technology must be a "a human response to and collaboration with God's own Word giving shape to time and materiality."[12] I share Brock's concern for keeping relationship with God central to the ethical evaluation of technology. However, his generic use of a singular category of humanity obscures the very specifics of collateral human damage present in both the scripture stories of Ark and Tower building. In other words, he doesn't question the role of language as a technology. I do find it important to recognize that Brock perceptively turns attention to the ways technologies shape and are shaped by cultural elements that create very specific power dynamics. Yet he does not unpack the possibilities of this turn related to Babel.

Similar to Brock, Quentin J. Schultze also uses the Babel narrative to critique human valuation of technology. In *Habits of the High-Tech Heart: Living Virtuously in the Information Age*, he turns to Jorge Luis Borges's "The Library of Babel" to confront what he calls "Database Babel."[13] Schultze likens the informational overload of digital technology, informationalism, to the chaos created at Babel. "In the digital future, our communities will reflect even more the confusion of Babel,"[14] he notes. He suggests that the creation of information and its sheer quantity are valued over and against the quality of the information and thus communication.[15] Similar to Brock, Schultze considers humans' being enamored with digital technology to be a false god. He writes, "From at least Babel forward we have churned through new techniques in the hope of making a name for ourselves. We tend to wear the clothing of technological power rather than the garments of personal humility."[16] Schultze argues for greater awareness on the part of users of technology. Yet he also has a rather pessimistic view of users' abilities or desire to engage technologies in a way that morally directs information use.

John Dyer, a theologian and web developer, joins in the cautions about using technology for redemptive capacities or fulfilling God's work with humanity. However, Dyer reads in the narratives of Ark and Tower building that God "values not just humanity but also the creations of humanity."[17] Specific to his interpretation of the story of Babel, Dyer notes that the tower builders were "abusing their creative powers."[18] He considers the fact that God could have leveled the Tower of Babel as God did the walls in Jericho. Instead, God chooses to confuse the languages. Dyer notes that "language is something that we create not only to communicate information but also to establish a sense of identity and inform the way we see the world."[19] Dyer concludes by summarizing his claims related to the Tower of Babel and technology: first, God also uses technology, not just humans; second, technology is part of the social world; and third, technological changes also impact culture. Thus, "God was essentially reprogramming their sense of self, their relational connections and how they viewed the world,"[20] writes Dyer.

Dyer's reading of the Tower of Babel and his related theological conclusions about the value of technology concentrate on aspects of the narrative that are different from Schultze's or Brock's. Dyer's reading resonates with important historical-critical elements raised by biblical scholar José Míguez-Bonino. Language is a form of technology that shapes cultures. God redirects the creative powers of those at Babel from imperial dominance of sameness toward difference and a required intentionality to understand each other. It is precisely attention to culture and differences—linguistic, geographic, and political—to which Míguez-Bonino calls the reader. He writes, "A new empire had reorganized life, redivided the land, reunified the former empire with a new center . . . *and a new language.*"[21] Previous interpretations of Babel assume a cohesive community with the intention to be better than God. A closer reading of the context of the text shows Babel was a land conquered by a "tyrant" (Gen 10:8–10), expanding the Assyrian-Babylonian empire (Gen 11:2–3).[22] The city was not a monolithic group of like-minded folks, gathered together plotting against God, or even a group that scholars can prove originally shared a single language. Rather, the ruler was enforcing a singular language and shifting cultural practices. Míguez-Bonino suggests the passage, then, is not primarily about the creation of diverse languages "but the condemnation and defeat of the imperial arrogance and universal domination represented by the symbol of Babylon."[23] If this is the case, then God's action "is twofold: the thwarting of the project of the false unity of domination and the liberation of the nations that possess their own places, languages, and families."[24]

The text presents a conflict between the universality of the empire and difference in God's creation. While the rulers at Babel try to create a monolithic, homogenized center, *"God's intention is a diverse humanity that can find its unity not in the domination of one city, one tower, or one language but in the 'blessing for all the families of the earth'*

(Genesis 12:3)."[25] Míguez-Bonino then invites the reader as he believes God intends to search for unity that does not sacrifice difference. As we consider the ways digital technology via algorithms may lead to unity without sacrificing difference, we must question those forms of "empire" that are working to limit difference. How does the limiting of difference via algorithms benefit corporations? In what ways does control of informational output direct social behavior, such as consumerism?

Christian scholars are not the only writers using the metaphor of Babel as they consider the impact of mass amounts of information and communication changes brought about by digital technology. In *Globalization and Media: Global Village of Babel*, Jack Lule reports that "Tech Giants on the other hand consider digital technology and electronic communication a way of regaining what was lost at Babel. They suggest both through translation and the ubiquity of common media that again one language is being created."[26] The goal of creating one language assumes that the confusion of the languages, or restoration of languages as Míguez-Bonino frames it, had a culturally and morally negative impact. As people seeking to value difference and expand (or at least preserve) the diversity of creation, we are correct in pushing back against the technology industry's goal or vision of a singular language that dominates and erases difference.

Keeping in mind the relationship between language, technology, and culture, Dyer and Míguez-Bonino provide a convincing argument about retaining the particularity of peoples, places, and languages as well as a cocreative role for humans in this process. Are algorithms bringing us closer to forms of communication that value difference or building a new Tower of Babel organized to benefit the empire of corporate capitalism through personalization of differences and reification of majority patterns? To explore this question, we need to better understand how algorithms are designed and used and how they may be changing us.

## DATA, ALGORITHMS, AND PREDICTIVE ANALYTICS

We have become so acclimated to the functionality of digital technologies that we rarely pull the curtain back on the wizardry of how things work. Journalist Chava Gourarie in "Investigating the Algorithms That

Govern Our Lives" writes, "As online users, we've become accustomed to the giant, invisible hands of Google, Facebook, and Amazon feeding our screens. We're surrounded by proprietary code like Twitter Trends, Google's autocomplete, Netflix recommendations, and OKCupid matches. It's how the internet churns."[27] Some respond with fear, and others marvel at the capacity of digital technology. The poles of moral panic and sacred reverence contribute to "a non-discerning, vacuous faith in the collection and dissemination of information as route to social progress and personal happiness."[28] We can and must have a better understanding of how digital information, algorithms, and predictive analytics work if we are to evaluate their moral impact on the digital world. This can be a difficult task because these three parts of algorithmic function are expansive and often unpredictable regardless of their naming.

There is the data—facts that are put into relationship with one another to create information. When this is done in a digital system, the data put into relationship creates digital information. The function of the algorithm is to determine the strength or weakness of digital information based on how the algorithm interprets the relationships that comprise the data linkages. Algorithms are designed to learn from each computational function and thus get increasingly better at recognizing and evaluating strength of informational value. This process allows algorithms to produce a prediction. The prediction or result is what we might call knowledge because knowledge comes from the practice of adapting or transforming information in a useful manner.[29]

Search engines are a perfect example of the various aspects of the digital data cycle. I type in a search term, such as chocolate chip cookie recipe. Before I've even completed typing the word *chocolate* (data), an algorithm is at work calculating the most popular searches that begin with the letters *choco* as well as aggregating my past search data from the same IP address, from my geographic area, and so on (searching for relational data points to evaluate the informational value of my search term). Faster than a human mind can work, the search engine provides auto-completed search terms, such as chocolate chip cookies, chocolate chip cookie recipes, chocolate chip cake, chocolate museum in NYC, and so on (predictive analytics). This process will be carried through in the site options I am provided and the priority in which they are presented. This is an example of the PageRank algorithm. The search data may then be integrated for future use and may even impact the advertisements I see over the next few days, utilizing a personalized marketing algorithm.

Through the simple description of a recipe search, we can dissect aspects of algorithmic function. What we do not know, and even the coders at Google may not know (assuming that was the search engine we used), is exactly how the algorithm worked. Did the auto-complete function list chocolate chip cookie before chocolate chip cake because more people search for cookies or because I have searched for cookies in the past? Or perhaps there is a cookie sale in my geographic area or my profile related to age, gender, race, family status, household income, and so on is more likely to eat cookies than cake? If for some reason there were a bias associated with cookies versus cake, this would have a more significant impact. Perhaps, instead, I was searching for images of people, and Google algorithms began to associate darker-skinned individuals with gorillas as happened in 2015 with Google's Photo app.[30] Algorithms are math or simple equations: "if A then B"; they make meaning out of information and predictions.[31] They serve as tools that can create solutions; they are not only instructions but also the implementation of those instructions. And they interact with each other in complex systems, creating according to their own directives with possible (even probable) unanticipated outcomes.[32]

Those who assume a digital technology functions on objective, rational processes see algorithms as clearing away human ambiguities or bias.[33] On the other hand, it has been proven time and again that human factors influence design and use as well as noting the bias in the outcomes or predictions made by algorithms. Algorithmic function is not a bias-free, objective function. That observation should raise significant moral questions especially given the entanglement and dependence of our lives with algorithms.

## Is Digital Information Fallible?

Information, generating the relationship between data points, and the ability to process it are taken as a priori goods in a digital world. It is not until an incident like facial recognition software identifying African Americans as gorillas in Google searches that we question the quality of digital information, the relationships that bring data points together.

Roberto Simanowski claims, "Software developers are the new utopians, and their only program for the world is programmability." He
suggests that programmers rely on algorithms as "the secret heroes of
this 'silent revolution' . . . that are taking over humanity."[34] As algorithms reshape cultural values and social norms, we must question what
responsibility programmers take for algorithmic outcomes and question
whether all data collection and informational computation are social
goods.[35] There is good reason to be careful in assessing responsibility
and accountability when identifying social harms that originate from
algorithmic function. Yes, algorithms are built by humans, but they also
learn on their own. The data algorithms rely upon is generated by
fallible humans; it is not value free.

For example, the suggestion of whom you might "follow" on Twitter
is generated by an algorithm. That algorithm has to account for a myriad of data points related to your current follow list; connections of who
follows you; and data points from your profile and usage, such as geographic location, educational status, and post content. Who you are and
the shape of your current follow list are data points from which the
algorithm delineates other possible connections.

Recently, much has been made in political discussions about the
diversity of social networks. The vast openness and accessibility of free
social networks leads people to believe they are egalitarian sites of diverse community. However, research related to social networks shows
that they are much more homogeneous than most of us would like to
admit. For example, recent findings in the General Values Survey by
the Public Religion Research Institute (PRRI) demonstrate that racial
segregation is perpetuated by social networks in the United States.[36]
The findings show that White Americans' social networks are 91 percent White and Black Americans' social networks are 83 percent Black.
Hispanic Americans have the most diverse racial social networks at 64
percent Hispanic, 19 percent White, and 9 percent some other race.
Moreover, 75 percent of Whites' social networks are completely White.
By comparison, 65 percent of Blacks have Black-only networks, and 46
percent of Hispanics have homogeneous racial social networks. Interestingly, when looking more closely at White social networks, homoge-
neity is not influenced by political affiliation, geographic location, or
gender and only slightly by age. The racial segregation we perpetuate
from in-person lives to online lives complicates the promise of open-

ness.[37] Segregated social networks are not only an algorithmic function. Social networks begin with humans who choose their list of friends or followers, often initially generated from personal address books.

As users show racial preference for friends and followers, the software then responds in that direction as well. In this case, users are the originators of the bias. Yet the programming preference to develop one's network based initially on an address book contributes to the problem. Admittedly, prioritizing other preferences such as geographic location or educational community may have the same results because much of the United States is racially segregated by geography and education. Scholars take different positions on the extent to which encoded values in social media platforms shape participation, but there is general agreement on a causal impact. Defining the technology as "open" sets the debate in dichotomous terms as open versus closed without considering what happens in the openness and who is served by it. In other words, users in participatory cultures (social networks) both bring and create hierarchies in the open space of the internet.

The vast openness of social media generates a variety of goods and harms related to equitable participation.[38] Social media is "open" to all in principle; however, the construction of many platforms is not egalitarian, and users bring hierarchical statuses with them. That is to say, there is a difference between access or interaction and participation. Participation is the key to social media engagement and network creation. Some describe this connection as a form of participatory culture that stems from user-generated content that combines cultural production and media sharing.[39] Participatory culture has often been limited to groups who come together based on a shared practice, including resistance to a structural power. The ubiquitous use of social media requires we consider how social media is, or contributes to, participatory culture. I will address this further in chapter 5, where we look at specific communities and their development of digital literacies. In this instance, the primary issue is how to distinguish between what humans are creating related to culture and what technology creates or generates.

Henry Jenkins, to whom the term *participatory media* is often ascribed, makes a distinction regarding the social shaping of technologies. He suggests that technologies are not participatory; rather, culture, created and organized by humans, is. He writes, "Technologies may be interactive in their design; they may facilitate many-to-many communications; they may be accessible and adaptable to multiple kinds of users; and they may encode certain values through their terms of use and through their interfaces."[40] For Jenkins, ultimately, it is the people operating the technologies who create culture through their participation. However, when people are unaware of the social-shaping influence of algorithms, they are unconsciously allowing the values of the technology to drive their digital lives. This is precisely why we need to increase our digital literacy. As users, we can live in diverse communities of difference or shrink our spheres of engagement through digital practices.

Other hierarchies related to labor are also perpetuated by the openness of digital platforms. In *The People's Platform*, Astra Taylor describes social media as digital feudalism or sharecropping where the platforms offer "open" land and then lay claim to content and mine data, shaping the labor and desires of users.[41] Such practices are having significant effects on the labor of those migrating to digital spaces, such as journalists, academics, and artists. Participation is enhanced by openness, access, and collaboration, but it also creates avenues for misuse of knowledge production and reduces if not eradicates compensation for labor. Andrew V. Edwards agrees in *Digital Is Destroying Everything*, warning that "where you are not paying for the product, you are the product.[42]

Increasingly, social critics argue that social media companies exploit the labor and social goods of users (aka free data) by promoting collaboration or sharing. Some platforms brand "sharing" or collaborative services as a social value, leading users to invest their capital and labor into a system that does not provide equal reward. Some argue, for example, that the benefits to people of connection and learning that come from their Facebook activities are a small price to pay for relinquishing their personal data. Scholar activists such as Laurel Ptak argue that we should lobby Facebook for compensation of "labor" via data generation: "They say it's friendship. We say it's unwaged work. With every like, chat, tag or poke our subjectivity turns them a profit. They call it sharing. We call it stealing. We've been bound by their terms of service far too long—it's time for our terms."[43] Wages for Facebook is her response. Another alternative to the monetary exploitation (technology companies make millions of dollars in profit) of participation (giving up data for free) would be to tax the companies behind such platforms and use the money to compensate "open access research as a public good"[44] or support funds for worker benefits, training, or cultural spaces such as libraries. Such lines of activism reflect historically rooted economic arguments to move from free trade to fair trade and raise awareness about workers' rights. This requires the technology companies to "pay back" into the system that keeps them going.

## Do Algorithms Engineer Sameness?

The value calculation of personal data is difficult to assess. Perhaps user data is a fair price to pay for access to social networks or various software services. We may be skeptical of the return on investment when we discover what little control we have over how our information is processed and then shapes our experience of those networks and programs. Most users of Facebook believe their news feeds reflect the material their "friends" post. They rarely question the algorithm that generates which posts they see or don't see. Facebook has determined that what you see should reflect what you "like." That means posts that are more popular (have more likes, emotional responses, or comments) are more likely to be ranked higher in a news feed. In addition to popularity, there are other factors that influence rank of posts in news feeds such as how many friends you have in common with the original post author, whether the content relates to a perspective you have liked in the past, the content's authenticity level, or perhaps how related the content is to a current event.[45] There are some friends' posts you will never see. There are others you may see all the time. Their appearance in your news feed is not a reliable indication of their personal connection to you or your valuation of the importance of information they are sharing. It is an algorithmic chance built on so many factors that it is difficult for even Facebook to tell you exactly why you see various posts over and against others. Regardless, the factors used generate bias related to personal popularity, political ideology, race, gender, and perhaps a host of other affinities. The algorithm exacerbates biases that are already part of your network and perhaps further shaping it in biased ways.

On dating sites, algorithmic bias is not only deepened based on user preferences but built right into the code. Marriage and dating are often referred to as social relationships with strong homophily—that is to say, we date and marry people with whom we have a majority of traits in common related to economics, race, geography, education, family background, religion, and so on. Of course, there are always exceptions to this rule, and we might think online dating sites are the exact place we can escape from these identifiers. Research has found that dating sites actually reduce your opportunity to meet diverse people rather than expand it. Molly Niesen's research suggests,

> Rather than connecting people based on commonalities, ODSAs [online dating sites and applications] have the potential to enhance differences, providing platforms for specific groups based on intersections of identity. More importantly, racism, sexism, homophobia and body shaming have become normalized on the world of ODSAs, inviting practical questions about if and how to make ODSAs more inclusive.[46]

The reassertion of homophily is also caused by the increase in sites that cater to specific identity-based categories, ranging from religious affiliation to race, urbanites to farmers, or more specific attributes like height or having a beard. Then, there are the platforms themselves. On eHarmony, for example, "although a 33-year-old man could be matched with a woman who was as young as 23, women of the same age could only be matched with men who were older than 27." Adding to age discrimination, lesbian, gay, or bisexually identified users were rejected by the site as were those married four or more times and those whose scores indicated depression.[47]

When algorithms reinforce bias based on socially constructed identity categories, we move further from both the net utopia original internet designers envisioned and the diversity of community affirmed as part of creation. The foundation of current social equality is based on "outdated and erroneous assumptions about the biological nature of identity."[48] This is evident when we look at the rise (and fall) of biometrics, the use of biological markers for identification technologies like fingerprint and retina scanners or facial recognition software. In her book *When Biometrics Fail*, Shoshana Amielle Magnet shows the various ways that biometric technology and the algorithms upon which they are built make assumptions based on sex, age, race, and so on, further deepening social inequalities. She details the ways in which biometrics often fail in their use on women, people of color, and those with disabilities. For example, facial recognition software was designed with Caucasian males in mind as the normative standard of a "face," thus preventing recognition of people with darker skin or proportionally different facial features. She shows how race, gender, and age affect fingerprint quality, making it more difficult for women of color and older individuals to have reliable fingerprint readings. Security firms that have invested in iris scanning have found the software is affected by eye shape and darkness of surrounding skin, making it more difficult to reliably read Asian and darker-skinned people's irises.

Since biometrics are increasingly used for security measures at airports, in prisons, and for social services, Magnet concludes:

> We have an impoverished language for thinking more broadly about such technological failure. We must also think of the intensification of existing inequalities as failures. For example, biometric technologies that rely upon erroneous assumptions about the biological nature of race, gender, and sexuality produce unbiometrifiable bodies, resulting in individuals who are denied their basic human rights to mobility, employment, food, and housing. Although biometric scientists often speak of "false accept" or "false reject" biometric errors, we lack language for thinking about the failures of biometric technologies to contribute to substantive equality.[49]

Magnet's evaluation of biometric design seems to have much in common with the architectural plans at Babel. A normative standard already based in social bias for one language of the ruling party is enforced upon everyone, co-opting the bodies and communication of everyone else in the vicinity. The architects in the case of biometrics are a mix of human designers and machine learning.

Algorithms are not objective processes free of bias, based on the data that they utilize and the way in which they are designed. Sorelle Friedler in "Being Hopeful about Algorithms" suggests that if we look for an "audit trail from an algorithmic decision maker we do have the hope of revealing implicit preferences and biases based on the algorithm's 'life experiences' aka 'training data.' And so we can expect more because we have the ability to do so."[50] That often requires beginning with the output of algorithms and then tracing back how the data could lead to particular outcomes or predictions.

## Are Predictive Analytics the Builders of Babel?

What Facebook, Twitter, or Instagram post we see or do not see may not seem like a significant social issue. Yet algorithmic bias can lead to dire consequences for some. The use of predictive analytics ranges from recommendations for shows on Netflix to personalized search engine results to sentencing determinations in courts of law. The use of algorithms for predictive purposes by the criminal justice and health care community have significant effects—some positive and some negative. More and more public health initiatives use hot-spotting, a form of predictive analytics to determine how and where to spend scarce educational resources, health services, and policing. Unlike biometrics that use biological markers as data points, hot-spotting uses social factors such as poverty, past abuse incidents, health care visits, and arrests.

For example, in Fort Worth, Texas, a local hospital uses predictive analytics that they call Risk Terrain Modeling to map and hopefully prevent child abuse. Doctors noticed a particularly high rate of child abuse cases in the emergency room as well as child deaths due to child abuse. Using information related to poverty, domestic violence incidence, and aggravated assault reports, the hospital was able to pinpoint neighborhoods, even specific blocks, where children were at high risk for child abuse. They reportedly caught 98 percent of future cases. The hospital combined educational and health awareness efforts as well as better detection efforts during emergency room visits.[51] No doubt this is a success when one judges safety of children as a moral good and integral to a good society.

The procedure also raises certain questions about how we detect child abuse. Domestic violence advocates counsel us to remember that violence does not discriminate based on race, geographic location, educational status, or economics. In other words, domestic violence is everywhere as is the specific form of it we know as child abuse. In wealthier communities, victims can afford private care or to travel to various doctors so no pattern is detected. Many of the data inputs to predict child abuse would not show any need for intervention in communities outside of poor neighborhoods that have disproportionate interaction with police and are often communities of color. Success like that in Texas and other locations around the United States has also led to a profit-making venture for organizations who sell predictive analytic software. In "Who Will Seize the Child Abuse Prediction Market?" Darian Woods details the money to be made and the lives saved by software such as Rapid Safety Feedback. He compares the use of predictive analytics with older forms of risk assessment such as Structured Decision Making (SDM). Woods reports, "This tool is a static snapshot; unlike predictive analytics it is not updated in real time. While not making use of real-time administrative data to the same extent as predictive analytics, one advantage of SDM is that it has been thoroughly researched."[52] It employs a wider range of data points that are cross racial, educational, and geographic markers. It also pairs training with local implementation, which requires an ongoing level of worker investment and discernment in the process.

Similarly, many criminal justice decisions across the United States are made based on predictive analytics. Algorithms are used to determine if a defendant is allowed to post bond or may inform sentencing if a defendant is convicted. Such systems are used to remove explicit and implicit bias on the part of judges. Recently, a ProPublica study of predictive analytic use in Broward County, Florida, showed racial bias. While the tool does not specifically ask about race, questions like "Was one of your parents ever sent to jail or prison?" and "How many of your friends/acquaintances are taking drugs illegally?," which ProPublica reports, "may be seen as being disproportionately impacting blacks." Also, the prediction of those forecasted to commit a violent crime in the next two years was correct only 20 percent of the time but was 61 percent accurate in predicting recidivism, which may also be directly affected by race. ProPublica "claimed the algorithm falsely flagged black defendants as future criminals, wrongly labeling them this way at almost twice the rate of white defendants."[53] For its part, the company that licenses the software reiterates that they do not use race as one of the data points. This reinforces the interpretation that the other markers are already racialized if there is a two-to-one difference in the results when race is not a stand-alone category.

Predictive analytics can be used to save some children from child abuse even if they are racially or geographically biased, though such biases also lead to incorrect sentencing and bond information for Black men in the criminal justice system. It is difficult not to see a correlation between the data points used in both models. Perhaps asking different questions is what's needed: What social inequalities lead to crime or how does antiviolence education in every school and faith community reshape cultural tolerance for gender-based and child violence? Are we really solving social problems or simply using predictive analytics to better pinpoint the factors that cause them? These examples highlight the fact that we are not disrupting and are in some cases reinforcing social oppressions through our use of predictive analytics.

In "Being Hopeful about Algorithms," computer scientist Sorelle Friedler reminds us:

> We've built up over the decades and centuries a system of checks and balances and accountability procedures for evaluating the quality of human decision making. We have laws that require non-discrimination, we have ways to remove decision-makers who make arbitrary decisions, and we have a social structure that makes decision-makers feel a sense of responsibility for their decisions.

None of these exist for algorithmic decision making or realistically can.[54]

Herein lies the structural and social responsibility associated with algorithms. Humans design the algorithms, create the data that is input, and often implement the outcomes or follow the predictions. Humans are as responsible for algorithmic bias as the powerful Babylonian rulers who came up with the idea of building the biggest tower by forcing a single language on the community of workers. Recall the biblical interpretation: many individuals were co-opted into the design and goal of oneness, perhaps some by force and others unconsciously.

Responding to predictive analytics on a social or structural level is made more difficult by their often unseen and at times positive effect on our daily lives. In a way, our everyday interaction with predictive analytics inoculates us to their graver social consequences. Without an algorithm, we would need to maintain connection by searching each individual friend's or follower's page to read new posts. We would have to use paper maps and listen to live radio for traffic updates. That is tedious. Because of our love of efficiency and overzealous trust in machine learning, we acquiesce.

Too often complacency and lack of awareness keep us from seeing that the data we generate is used in a feedback loop to promote homophily and aggregate out difference, making it easier for platforms to shape our desires as well as profit from our consumption. As Eli Pariser writes in *The Filter Bubble*, "More and more, your computer monitor is a kind of one-way mirror, reflecting your own interests while algorithmic observers watch what you click . . . shaping via personalization, what you see and what you want."[55] Users initially believe that they are *pulling* media by curating friends, promoting tweets, or googling research questions. However, users are more often than not getting *pushed* particular data that is popular among their friends or that Google has determined best aligns with their past searches. Filters lead to a personalization that promotes confirmation bias of our own opinion, thus decreasing all kinds of diversity in our networks, which as we know are already significantly racially homogeneous.[56] Essentially, we are not only fighting structural examples of Tower building, but we are also constructing a Tower of One. If we are inside our Tower of One, it is increasingly difficult to identify the more powerful tower-building forces. It is time we recognize that we rarely curate our own information or direct our own search; we need to break out of our towers, scatter our languages, and start searching for differences.[57] Alternatives abound related to both personal and structural change.

## SEARCHING DIFFERENCE, NETWORKING DIVERSITY

God's active reengineering of difference at Babel is a reminder of the value of diversity brought about by the particularities of languages that build cultures and histories. No group is given dominant control, and no hierarchy of difference is established. Humanity throughout history has struggled to create relationships and community that exemplify diversity. If God reinforces difference as a means to preventing domination and oppression, then we need a system of community that promotes an understanding of differences, not an erasure of them. In our relationships with each other and with digital technologies, we are called to break open technological structures and behavioral practices that confine and separate people.

Ethical response to the examples raised in this chapter requires flexibility and creativity. Those with various technological skills will bear greater responsibility because they have the capacity to alter algorithms at a variety of stages. In addition, all of us can respond first by developing an awareness of the function of algorithms, second with a disruption of default settings, and last by questioning predictive outputs; these are key aspects of digital literacy. A Christian ethic that values growth and responsiveness is key to meeting the moral challenges presented by algorithmic bias. As algorithms learn and change on their own, we will need to tease out the role humans have played in their trajectory and reorient algorithmic development, use, and ongoing responses in ways that interactively reduce bias.

There are a number of ways as users and programmers we can step into an active cocreative role with our technologies to magnifying diversity rather than reducing it. The first step is to demystify digital technology as a god in and of itself. Samuel Arbesman writes, "technology, while omnipresent, is not pristine or unfathomable because of its creation by some perfect, infinite mind. It is wonderfully messy and imperfect. And it is still approachable."[58] Granted, we cannot know all the ways an algorithm functions or even come close as human beings to mimicking its computational power.[59] By scrutinizing our data inputs and following the trail of consequences related to outcomes and predictions, we can begin to assess the moral impact of various aspects of algorithmic function.

Users, programmers, policy makers, and tech companies all have a role to play in disrupting algorithmic bias. For users, the first step is simply to recognize algorithms at work in our everyday use of digital technology. We do not need to know the full complexity of how each algorithmic aspect works, but we do need to understand the basic structures. This chapter has sought to provide some of that information as it relates to issues of social inequality and promotion of difference. There is more that can be learned. Microsoft Social Media Collective has created an online bibliography "in an attempt to collect and categorize a growing critical literature on algorithms as social concerns."[60]

To pop our own filter bubbles or Towers of One, we will need to be creative. The basic awareness and understanding of algorithms allows users to see when they are promoting difference or erasing it. Roberto Simanowski in *Data Love* suggests sabotage—adding multiple users to singular accounts to disrupt coherent profiling or using apps that click every like on a page to disrupt prioritization by popularity—or boycott, which may mean using privacy settings, regularly clearing browser history, and being selective about or leaving social networks all together.[61] Users, like myself, who are enmeshed in Google products can turn off their Google search personalization or switch to a different, private browser platform. When it comes to civic engagement, we can intentionally randomize news sources or read more than one source for the headline stories. Eli Pariser calls this "taking a new route" via internet travels. He writes, "Habits are hard to break. But just as you notice more about the place you live when you take a new route to work, varying your path online dramatically increases your likelihood of encountering new ideas and people."[62]

Users' biological and social differences, such as race, sex, gender, class, education, ability, or health status, require different approaches to disrupting algorithmic bias because some of us are the beneficiaries and others the recipients of such bias. For White individuals, who benefit from the pernicious social oppression of racism, we have a particular responsibility, given the overwhelming data on racial bias in algorithms.[63] Given the normative default setting of whiteness in our lives and in software design and use, White people, especially in the United States, need to take radically different routes home. That may mean visiting and learning from sites such as "Racialicious, The Root, Colorlines, The Asia Society, Ill Doctrine, NewBlackMan, MuslimahMediaWatch," and the Microaggressions Project.[64] It also requires an intentional diversification of our social networks in terms of friends and followers.

Joy Buolamwini, founder of Algorithmic Justice League, is on a mission to eliminate algorithmic bias in favor of justice practices. Her project is one that *breaks open structures that confine people*. In exposing the racialized coded gaze, she is working to fight bias. The Algorithmic Justice League has information and engagement strategies for citizens (demanding fairness, accountability, and transparency), activists (mobilize for change), artists (provoke reflection), academics (research bias), companies (check for bias), legislators (change policies), regulators (create standards), and coders (inclusively develop). Buolamwini is starting a movement for "incoding," a process of coding that begins by asking "who is missing?" to disrupt algorithmic bias at its beginning stages.[65] The notion of coding with particular values in mind disrupts the claim that computational science is a value-free, objective endeavor. There are always values involved, which means we can be intentional about valuing differences rather than dominance.

In addition to technologists like Joy Buolamwini, many others are working to bring about greater diversity and combat bias. Interventions take place at all stages of the algorithmic function; a group of Canadian computer scientists have developed a mechanism to certify and remove disparate impact[66] of algorithms. This group of researchers used values given to input data to determine the level of disparate impact (or bias) of an algorithm. Because access to the processes of the algorithm are often unattainable, they focused on repair procedures for data sets where they found disparate impact. For those readers with a greater level of technical knowledge, the scientists used numerical attributes, often having to recalibrate data sets, but claim the processes could also be expanded to categorical data or vector-valued attributes.

Other programmers are bridging the computer learning divide by paying closer attention to language. National Public Radio reported on the work of Adam Kalai, a researcher at Microsoft who is working to make computers less sexist. He starts "with the bits of code that teach computers how to process language . . . something called a word embedding. . . . Word embeddings are algorithms that translate the relationships between words into numbers so that a computer can work with them."[67] As Míguez-Bonino reminds us, language shapes culture, is informed by particular histories, and sustains the difference as part of God's creation. However, when computers transform the relationship between words from one language into computational forms of number sequences, they also absorb the biases present in that language. When running a test on what words were associated with *he* versus *she*, a word embedding dictionary returned *he = brilliant* and *she = lovely*, and more troubling, *he = computer programmer* and *she = homemaker*. As the story suggests, if one were using this program to sort through résumés at a tech company, the problematic gender associations would bump to the bottom résumés by women because their names and use of language would not be as closely associated with computer programmer as those that had male identifiers. Other stereotypical markers of race, class, and ability have also been found. I would guess religious bias would also be present. The dictionaries can be taught to ignore certain word relationships, which then results in a ripple effect of disassociation with other words. This work takes fighting bias and deepening differences to a level not recognized in our scripture stories. It is the work of *breaking open the very structures* of language itself.

Algorithms are neither the lone solution nor the only cause of digital bias and the shrinking of differences. They often function in a manner that following their human designers and users reinscribes the biases and hierarchies of our communities and personal relationships. Through more fluent digital literacy, we are the cocreators who are called to maintain the scattering of languages and work toward use and design of algorithms that increase our hearing and understanding of difference. As we hold onto the need to program for difference rather than dominance, we must also ask how being networked selves changes how we define the self and form relationships across differences.

In chapter 2, we explore Christian ethics resources related to virtues and moral formation that respond to rapid digital change.

## EXCURSUS 1: DIFFERENCE AND SELF-UNDERSTANDING

> And at this sound the crowd gathered and was bewildered, because each one heard them speaking in the native language of each. Amazed and astonished, they asked, "Are not all these who are speaking Galileans? And how is it that we hear, each of us, in our own native language?
>
> —Acts 2:6–8, NRSV

I grew up a cradle Roman Catholic. My family and friends were all Catholic. I went to Catholic schools. To be honest, I didn't experience other religious traditions until I went to seminary. I was challenged in the classroom by questions about my own and others' doctrines and religious practices. I learned about Protestant traditions; Judaism; and a small bit about Islam, Hinduism, and Buddhism. I encountered radically different experiences of Catholicism across racial, geographic, and historical streams. Most important, the diversity and questioning in seminary made me understand myself and better articulate my own beliefs.

Some Christians may wonder why understanding or engaging with different faiths matters. Pentecost, the story of the Holy Spirit descending on the disciples, is often told as an antidote to the Tower of Babel passage (explored in depth in chapter 1). God scatters the languages at Babel; at Pentecost, everyone can hear each other. This is often misinterpreted as there being "one language" again. Pentecost is not a Christian correction of Babel. That is a misreading and an example of Christian hegemony. Both biblical stories contain a similar message. Diversity, the human array of differences, is restored and reinforced by God. The text, "We hear, each of us, in our own language," affirms the diversity of the believers gathered. There was no need for uniformity or sameness.

One might think the internet provides such an encounter for 2.0 seekers and believers. Anyone can watch a video of just about any religious ceremony or worship (sometimes live streamed, allowing participation as an attendant) or join different forums or read a blog to get a sense of how others live their beliefs. The internet can teach us about different rules, doctrines, and historical shifts between our faith traditions and another's (look no further than Wikipedia). Unfortunately, the internet does not provide a genuine ecumenical or interfaith experience unless one works very hard at it. Comparing the diversity of what the internet offers with what users actually experience, we find that most of us live in very closed online spaces regarding religion, news, geographic location, and so on.

What do we lose when the richness of our diversity, whether religious, racial and ethnic, gender or geographic, is weeded out of our lives? One outcome is often that we become too certain of our own viewpoint. At the end of *Christian Ethics: A Historical Introduction*, Philip Wogaman writes, "too much certainty about God's ways with humanity, may not leave enough space for God to be God."[68] When we limit our experiences to people who look like us, talk like us, or even believe like us, we may be limiting the ways we can encounter God.

Both the variety of religious experiences and the diversity of my online communities enrich my faith. What can you do to deepen the diversity of your experiences? What types of diversity are missing from your social networks?

# 2

# NETWORKED SELVES

**D**igital technology has transformed the way we connect with one another, shape our identity, and form relationships. Many Christians are worried that digital technologies are replacing God's presence in our lives with techno-demigods. The concern is both theologically and technologically rooted. Christians have often struggled to define their relationship with God. Historically, God has been seen from a hierarchical position over and against matters of the world. To focus on one's relationship with God, Christians were taught to resist earthly distractions. The theological assumption is that God is not present in earthly things. Perpetuating this theological concern, digital technology is seen as constantly grabbing our attention for trivial, human matters. At the same time, digital technologies are acting more and more human in ways we once thought only God could create.

Counter to these interpretations, we are equally able to recount the effects of digital technology in a way that envisions relationship with self, other, and God in a more generative and responsive manner. We know from our own creation story that humans carry the image of God and that all of creation is the signature of God's presence and intentions. Person-to-person relationships can make God present among us. Nurturing God's creation brings forth the creation God established. We can no longer participate in our world apart from digital technologies; they are part of who we are in different proportions, dependent on economics and geography, from global interfaces to an individual user.

In all human actions, there is moral meaning for our relationship with God. Digital technologies heighten a self-understanding as "networked," or relational. In an analog past, we might have discussed how God created us as relational beings, connected to one another, our actions impacting more than the self or the immediate persons we know. Yet we could not experience connectivity in the same way we do via social networks. The ripple effects of our connectedness are tangible and quantifiable in a way they were not before. Thus, a networked understanding of the self is highlighted as a key aspect of digital literacy.

In fact, our sense of self as relational leans into the Christian trinitarian view of God as three in one, relational even in God's beingness. Our digital experiences revise the traditional view of God as separate from us in a hierarchically directed "up above" theology. The metaphor of God as Wi-Fi or the internet is too instrumental, though middle school youth understand the omnipresence of God much better with these metaphors than any others I have used! What I mean is that digital technologies provide an experience of relationality unprecedented in previous generations. This is not always good, and I will address that in this chapter. Social networking or online participatory behavior highlights patterns of moral formation to which Christians must attend. The patterns also evidence the ways in which technology and human behavior interact to shape each other in coconstitutive ways. In this chapter, we will focus on two aspects of how digital technologies shape "who we are" as "datafied" selves and the impact these experiences have on our theological understanding of self, other, and God.

The immense creative power of social networks or online participatory platforms opens up spaces that are embodied yet beyond geographic location, time dependence, and fleshly limitations. Using a virtual-reality viewer, one can travel across the globe to see the sunrise around the world. Viewing these magnificent scenes, the body responds to the sounds and sites (and soon with other technological advances, i.e., smells). Joining a virtual community allows users to design an avatar that may look just like them or nothing like them and interact in everyday activities involving conversation through shared audio or typing. Some platforms allow for experiences we may never do in person, such as sexual encounters with strangers, kingdom building in ancient times, or modern-day warfare. While the avatar may not be the "real" person, it is an aspect of the user's self that informs who they are. Whether our relational experiences are online or in person, they inform the sense of self. Online networks allow for a wider diversity and expression of self than our fleshly bodies can accommodate. Simultaneously, the structure of these networks may also regulate and confine our sense of self and redefine relationships (e.g., "friend").

Second, we will address how digital technologies are increasingly and often seamlessly enmeshed with our daily existence. Some researchers suggest that we have already reached a cyborg-like status where we cannot exist unassisted by or disintegrated with our digital technologies. Others go so far as to suggest that "we have become data" and this is the only way we are intelligible to the computational world around us. Even examples such as the elderly or impoverished individuals who cannot afford personal digital technologies are intelligible to the world around them via technological systems. Every doctor's visit is logged in a cloud-based online system; tax, birth, and death records are all kept via searchable digital technology. The use of a credit card, ATM, or even a store transaction in cash is logged in a digital record-keeping system that marks where you were, what you bought, and when. Many people, even those in the global South, for whom we in the North may assume there is less access and connectivity to the internet, live digitized lives. Many pay large percentages of their income to keep a mobile phone connection and use wireless technology to connect to the internet wherever possible. That is only personal use; it does not capture the way in which global economic and political systems use data to track, inform, and predict, ultimately transforming how society, from local communities to nation-states, is defined. The society-wide impact of dataveillance will be addressed in more depth in chapter 3.

Users see and feel a change in relationship with our devices via push technologies—software designed to proactively reach out to the user rather than interaction being user generated, such as mobile application notifications for incoming email, recent friend posts, or buzzing of a Fitbit to remind us to walk. Push technologies create a dialogue or communication path between us and our digital technologies—one that can shift internal, self-regulation of attention and desire. Social science literature raises questions about adult internet addictions, the negative habits children are forming with so much screen time, or simply frustration at constant interruptions during everyday conversations.

As an ethicist, I see these as questions of moral formation centrally concerned with our sense of "who we ought to be" and "what we ought to do." As Christians, we ask these two questions in relationship to who God calls us to *be* and what God calls us to *do*. The formation of self and our relationships in a digital world then relates directly to Christian concerns about virtue and how we can be most responsive to God in relationship with us. As noted above, our theological understandings must also be refined to reflect our experiences of God, self, and other. Often, we turn to a list of rules that we hope creates balance between faith commitment and technological commitments—no use of mobile phones in church, only two hours a day of screen time for kids, never post negative comments or pictures on social media, and so on. We need guidelines to help us navigate our digital lives. Yet these responses still treat digital technology as a tool we can pick up and put down. The integration of digital technology with the self is no longer a separate tool; it is a way of being in the world. We need a richer understanding of digital technology, a literacy with the datafication of the self, to approach moral formation of self-in-relationship that honors our embodiedness and considers how we relate beyond bodily limitations.

In this chapter, I explore Christian ethics writings on attunement as a virtue that guides digital living. The increasing influence of digital technology reshapes our sense of self in ways that may lead to greater connection or make us feel disconnected. Our networked sense of self requires an ethic of digital literacy that includes consideration of impression management, the curation of online self-presentation, because it impacts who we are in relationship to God and others. Even for those individuals who are not on social media platforms, digital technologies define our existence, from health care records to government information and banking to the function of energy infrastructures. Attunement helps us orient ourselves as datafied, embodied, and spiritual beings.

## MORAL FORMATION IN A DIGITAL LIFE

Christian theologians describe virtue as the desire for the good,[1] or we might also think of virtue as characteristics of a person that are morally praiseworthy. Moral formation or virtuous living is a dynamic process with multiple factors that account for individual, communal, and structural forces and involve intuitive, learned, and creative aspects. Thus, we might think of virtues as social skills, "to have a virtue is to have extended and refined one's abilities to perceive morally-relevant information so that one is fully responsive to the local sociomoral context."[2] That is to say there is a particularity to one's virtuousness. Moral virtues are cultivated based on a number of factors, some of which may be innate, cultural, or interpersonal. In their essay "The Moral Mind," Jonathan Haidt and Craig Joseph elaborate on the interconnections between these factors.[3] For the purpose of discussing moral formation and digital technology, we do not need to tease out what is innate, cultural, or interpersonal. Rather, we need to be aware that they are all active in moral formation. Haidt and Joseph write, "For those who emphasize the importance of virtues in moral functioning, then, moral maturity is a matter of achieving a comprehensive attunement to the world, a set of highly sophisticated sensitivities."[4] This type of approach to virtue focuses on a dynamic link between practice, habit, reflection, and intuition rather than strictly abstract reasoning or top-down, memorized knowledge.

Attunement alters a historically popular orientation to virtue as seeking orderedness (following rules) or temperance of negative or lesser goods (restraining our baser desires) to an embodied and emotionally aware, even relational, approach to moral formation. Christian ethicist Cristina Traina writes about attunement related to the erotic, or desire for connection, in human relationships. In particular, she deals with relationships that involve unequal power dynamics, such as the relationship between parents and children. Traina says that attunement is "perceptive attention and adjustment to feelings, needs, and desires—both one's own and others.'"[5] The ongoing nature of attunement, which one might get better at but never master, is relational and responsive rather than based on behavioral goals. Traina is not interested in reining in desires or ordering them in a bad, better, best manner. Rather, she wants us to acknowledge desire and attune ourselves from a wholly and perhaps holy realistic and aware disposition. Erotic attunement, for Traina, or attunement as I am using it, is not a relativistic ethical stance that allows for "anything goes."[6] Attunement requires a reciprocal back-and-forth awareness of one's self and the other or others. Attunement, to erotic love, as Traina describes it, requires practice because it "combines perception, imagination, and experimentation in an endless, partnered dance"—a dance that partakes in a self-correcting process.[7] Attunement focuses our attention on the process of moral growth in a responsive, accountable, and expansive manner.

Our digital existence shapes not just our everyday mundane actions but also our moral sensibilities. "Every human act is a moral act. The way we talk, the time we spend, the plans we make, the relationships we develop all constitute the moral life. Morals is not primarily the study of grave actions; rather it is the study of human living,"[8] writes Christian ethicist James Keenan in *Virtues for Ordinary Christians*. He says that out of the complexity of life, we form particular practices and habits and as we face new experiences or interactions we require an "appreciative self-knowledge" for "moral growth."[9] This growth, what I am terming *attunement*, is not a lone, individual process but one that is relational with others and directly influenced by our relationship with God. Keenan argues that we need help to see ourselves as better than we currently are to grow in morally attuned ways. It is not simply about avoiding sins but positive development of self and relationships. In his final chapter, titled "Moral Virtues and Imagination," he describes how virtues help us see what can be, who we can be, and what the world might be and that this requires a sense of creativity and imagination.[10]

As Christians, relationship with God plays a central role in the moral imagination and moral vision we bring to our everyday lives. Earlier in the chapter, I noted how trinitarian understandings of God could support a networked and participatory approach that promotes inclusivity and provides innovative ways to *be* Christian. Dwight J. Friesen, a practical theologian and author of *Thy Kingdom Connected: What the Church Can Learn from Facebook, the Internet, and Other Networks*, discusses how "the uniquely Christian understanding of God as triune paradoxically draws 'otherness' together in oneness. It is this kind of differentiated unity that we seek."[11] Friesen outlines how the theological understanding of God as triune implies that humans as created in the image of God reflect selves-in-relationship or an existence of mutual interdependency. In today's digital world, people as networked selves evidence this.

Reflecting on social networks, Friesen says links are like relationships and nodes connote the networked person. Networked people are not just made up of themselves but include their relationality to others.[12] He invites the reader to consider the way that social networks show us the intimate and infinite connection that humanity is. He writes, "If we are truly interconnected, then the existence of every person, whether we will ever benefit personally from them or not, contributes to the complex fabric of the human experience."[13] Thus, we need difference in our networks for them to actually reflect what God has created and to better know ourselves. Friesen redefines the kingdom of God as an "open *We*." The moral virtue of attunement, using Friesen's imagery then, reflects and strives for the "open *We*": as "we proactively seek to help life flourish while also proactively standing against injustice and the oppression of life, we embody the 'open *We*' of God."[14]

## ARE WE DISCONNECTED IN OUR CONNECTION?

Digital technologies connect us. We exist as networked selves in unprecedented ways. Do these connections result in a sense of interdependence, or are we increasingly disconnected in our connection? What of our own actions and perception of connection via digital technology and, specifically, social media? In *The People's Platform*, Astra Taylor contrasts two dichotomous reactions to online participatory platforms.[15] She suggests that supporters of social media often attribute to it the power to liberate humans, expand our imaginations, develop never before seen communities, and make us better citizens. Detractors consider social media to have ensnared us in virtual chains, dulled our senses, increased isolation, and shaped us into more efficient consumers. Both have some truth to them, but neither is completely correct. In such debates, it is common to isolate the users and the technology, putting the blame on one or the other. Social media creates a space where humans become produsers (producer + user) and prosumers (producer + consumer) of technology and information more generally.[16] That is to say, technology does not determine the user but the user's participation is not free from being transformed by the values and purposes for which the technology is designed.[17]

Technological revolutions related to broadband access, mobile devices, and social media platforms have significantly shifted access to information, forms of communication, and divisions between private and public as well as human networking.[18] These shifts impact relationship formation. In the introduction to their volume *Digital Media, Social Media and Culture: Perspectives, Practices and Futures*, Pauline Hope Cheong and Charles Ess suggest, "Very clearly, digital media facilitate and mediate social relations, including people's notions of relationship, patterns of belonging, and community."[19] We are relational people; we want connection whether via text message, phone, Facebook, or Pinterest. Media scholar Lisa Nakamura notes that social media produces the "desire to connect and the need to self-regulate."[20] In particular, she notes that women's self-regulation on social media is often in response to unwanted and unwelcome misogynist behavior by men. For women of color, the response can be particularly virulent not only in its misogyny but racist as well. What information we share and how we shape our online profiles and interactions require regulation of privacy settings, appearance in photos, type of language used, and so on. These forms of regulation are often set by the platforms we use, though not solely.

Sometimes, our experience in these webs of relationship leave us feeling isolated rather than connected. In the well-known book *Alone Together: Why We Expect More from Technology and Less from Each Other*, social scientist Sherri Turkle suggests that digital technology is actually instrumentalizing human relationships rather than deepening them. She writes, "The self shaped in a world of rapid response measures success by calls made, e-mails answered, texts replied to, contacts reached. This self is calibrated on the basis of what technology proposes, by what it makes easy."[21] There is a loss of uninterrupted thought, self-reflection, and just being. Turkle describes a life defined by the computer paradigm or based solely in the values of the digital world—speed, quantity, profit, and efficiency.

As long as we allow the values of digital technology to drive moral formation of self, we may be more networked but less relational. "In a surprising number of epistemological traditions, introspection is a key ingredient of informed decision-making. However, introspection requires time,"[22] writes media theorist Kerri Harvey in *Eden Online: Reinventing Humanity in a Technological Universe*. Computer-mediated communication most often pushes instantaneous replies. For this reason, she warns, "More and more, it is on technological bones that human self-definition is hung."[23] We need to be more aware of both how digital technologies shape desire for response and how they influence the way we define and value relationships—this is the core work of attunement.

## DIGITALLY CREATING THE SELF

Formation of self online is often discussed as impression management. We engage in impression management all the time; consider the choices we make about how we will dress when going to work, school, or church. Online platforms add new rules of participation related to impression management that affect self-formation and presentation. Many Christians scoff at the idea of translating secular ideas about branding or marketing for faith communities. Also, we are skeptical about narratives that suggest we shape our self-presentation as part of a "culture" of media. I have certainly been guilty of dismissing mega-church Christian pastors for spending more time on glitz and glamour than theological content.

For some, any talk of branding is anti-Christian and too capitalist to gain a hearing. Phil Cooke, a media executive and Christian, addresses this issue in *Unique: Telling Your Story in the Age of Brands and Social Media*. Cooke breaks down the basic concepts in a way that reminds us of the basic social and cultural practices in which we engage daily. We are constantly "presenting" ourselves to others. Presumably, we want our Christian identity and story to be known based on our self-presentation either by what we say, wear, or do in the world. Any form of self-presentation reflexively informs who we are as moral people in this world. That process is not solely owned by each individual. Cultural, historical, and geographic location shapes this as does our gender, race, ethnicity, body shape, socioeconomic status, physical and mental abilities, and so on. We try to present ourselves in particular ways, and we are also read by others based on their own experiences, knowledge, and assumptions.

Cooke suggests that a brand or your brand is essentially a story. He asks, "what do people think about when they think of you?"[24] He does not want Christians to shy away from the awareness of branding, especially in a digital media environment. Cooke argues, "stories have remarkable power, which is exactly the reason Jesus used them."[25] He continues, "Stories drill deeply into your brain and explode later with meaning. Sometimes the meaning comes when you least expect it. Stories impact audiences because each person interprets the story in light of his or her own personal situation and experience."[26] Narrative or stories help us make sense of moral issues; this is also the main purpose of Jesus' parables and stories. Haidt and Joseph, whose work I discussed above related to virtue and moral formation, remind us that narrative thinking or storytelling is an innate aspect of human cognition. They suggest that "human morality and the human capacity for narrativity have co-evolved, mutually reinforcing one another."[27] The link between narrative and morality is a cultural tool for modifying and socializing humans. "The telling of stories is an indispensable part of moral education in every culture, and even adult moral discourse frequently reverts to appeals to narratives as a means of claiming authority,"[28] conclude Haidt and Joseph.

Many readers will be familiar with the famous phrase, "The medium is the message!" In 1964, Marshall McLuhan, a media theorist, published a now seminal text in media studies, *Understanding Media: The Extensions of Man*, in which his first chapter is titled "The Medium Is the Message."[29] Much of what he said in that text and in the subsequent lay version of the text titled *The Medium Is the Massage* (1967) (based on a typesetting error by the publisher but kept by McLuhan for its multiple meanings, like mass/age or massage as it relates to slight forms of manipulation) foreshadows and predicts issues that have come to fruition in digital communication. He argues for a shift in attention from content to stressing the importance of the medium. For example, consider the different experiences produced by hearing scripture read from the pulpit, reading a scripture text printed in a bible, or reading the specific verses on a bible app on a smartphone. The medium of speech, written text within a book, and segmented electronic text produce different experiences even if the content is the same. McLuhan's work forces us to deal with the unknown effects shifting mediums have on culture, individuals, and social systems. There is much more that can be said about McLuhan's work. For our purposes, this main point generates plenty of insight and leads us back to considering how the medium of digital technology, specifically social networking, shapes the story we tell about ourselves and thus how we are formed as moral people in today's world.

Computer-mediated communication affects human interaction beyond issues already mentioned, such as speed. It raises questions of authenticity or coherence between online and offline selves. It also shifts how and with whom we interact, including issues related to audience such as control of perception, constituent contributions, and influence of dominant social systems. All of this is negotiated within the constructs of the medium, as McLuhan reminds us. We are, as Cooke suggests, always telling a story about who we are as Christians, which is an outward, though not always conscious, embodiment of our moral character. How is that story "read" differently if I am telling it via Facebook, a blog, or in person as we serve food at a homeless shelter?

On Facebook, I may post a picture, link to the shelter, and add a verse from Matthew 25. A friend can then investigate the rest of my page (presumably years of posting) to see if other aspects of my posting line up with this self-presentation. On a blog, I have more space to add photos and detail how and why I came to serve at this shelter. This medium allows more personal control over the narrative. Again, a visitor to my site may read other posts or follow links I include that provide data on homelessness and ways to get involved, signaling a different form of engagement. In person of course, I share with those present the immediate experience of faith in action, but they have little access to know more about my motivation for service other than what I choose to verbally communicate. It is not the case that computer-mediated communication provides less information or requires a fast read. Social networking platforms may in fact require more time and care dedicated to impression management, or curation of one's self-presentation.

Sociologist Erving Goffmann describes impression management in *The Performance of the Self in Everyday Life*.[30] His 1959 work predates the internet and social media; however, his description of self-presentation is perhaps more acutely felt in these environments where dramaturgical elements, such as front stage, back stage, audience, role, props, cast, and so on, are more easily identified. Goffmann was interested as a sociologist in describing how people negotiated performance and its everyday effects. Here, I am more interested in how awareness of this performance via social media shapes our ethical character. In other words, how is impression management part of moral formation?

Whether one is seeking a coherency or multiplicity of online and offline identities, all of these performances shape the totality of the self.

A key aspect of moral formation relies on integrity, the sense that our lives hang together. We are who we say we are, and we act in a way commensurate with that self-identification. Early internet pioneers heralded the bodiless space on online platforms to free users from the bodily restrictions that accompany our offline existence. For example, one could enter a platform and be a different race, ethnicity, or gender or leave behind a physical disability. Anonymity provided a creative space to *be*, beyond the oppressive identity elements of offline lives. Users often found it was difficult to leave behind the speech or interactive markers these identities had shaped, and thus being a different person online did not always work as seamlessly as one hoped. Additionally, many use the possible anonymity of the internet to act in violent, racist, sexist, and homophobic ways toward other users. Freedom of identification with one's offline persona did not lead to a liberated and oppression-free online world.

While early internet users may have created radically new presentations of self, Judith E. Rosenbaum, Benjamin K. Johnson, Peter A. Stepman, and Koos C. M. Nuijten have found that the Facebook effect of "real names"—requiring users to verify themselves and their association to their "real" name—has changed the way users interact on social media. In "'Looking the Part' and 'Staying True': Balancing Impression Management on Facebook," these researchers found "evidence is accumulating that online self-presentation requires a healthy dose of authenticity, or at least a balance of self-promotion and accuracy, and that Facebook profiles better match the user's actual personalities than their idealized selves."[31] They found that users prioritize goals in their self-presentation, not a new phenomenon in social interactions. In their study, Black college students in the southeastern part of the United States had three main goals: (1) creating an authentic self-presentation; (2) creating a professional, positive, and current self-presentation; and (3) controlling information, including what others put on their page. Unlike offline self-presentation, the third goal requires a different set of social skills. "In conclusion, our findings suggest that interaction is an important part of self-presentation, and just as vital as the construction of one's image, which appears to be a balancing act between enhancement and vulnerability, and between authenticity and selectivity, all of which is complicated by audience heterogeneity,"[32] find the researchers.

Participatory social platforms raise particular questions for impression management that are perhaps exacerbated by the functions of social media. For example, we have always judged people not just by their own actions but also by the crowd with whom they hang out, the people they associate with. This was a particular problem for Jesus. He was often with the outcasts and powerless of society—beggars, lepers, disabled, children, and thieves. Thus, Christians might have a different orientation to this aspect of self-presentation, though I doubt we are much different than Jesus' critics when we judge daily interactions with friends, neighbors, or strangers. Online, however, various platforms not only allow for association with certain groups but also allow individuals and groups to contribute to one's self-presentation by adding posts to one's wall, tagging related to a particular set of information, or simply posting a photo or video without permission that presents the user in a positive or negative light. The ability for this information to travel and be ever present is a function of digital media, unlike past analog ages when physical presence was the only way to know with whom someone associated. "The power to identify or to self-identify ultimately raises the question of to whom one's identity belongs,"[33] writes Bruce E. Drushel in "Virtual Closets: Strategic Identity Construction and Social Media." He argues that identity is now, more than ever, a multiplicity of negotiations that include a networked audience. That is to say, our identities are a relational construction rather than an autonomously owned action.[34] They also require an audience to be "realized," or made real.[35]

These aspects of social media and impression management generate a good deal of anxiety, especially for young people. They also require a significant amount of time to manage. While most research confirms users are seeking authenticity in self-presentation, they are also savvy about highlighting particular aspects that make them look better. In *The Happiness Effect: How Social Media Is Driving a Generation to Appear Perfect at Any Cost*, Donna Freitas chronicles the use of social media through the experiences of college students who represent those immersed in digital technology at the cusp of generational change from Millennial to Generation Z.[36] Most students with whom she speaks are unhappy with the overarching social pressure to appear happy and the feeling that if they are not present on social media, they do not exist. I raise this point specifically as a generational shift but also one that reflects a way of "being" that is not far off for most of us. We communicate less with friends and family who are not reachable through social networks or computer-mediated communication. Imagine individuals in the United States who use only a landline and can be reached only if they are physically at that location. Accessibility and connection to that person are drastically reduced. Now, consider a generation that primarily communicates via social media. If you are not on the platforms, you are not in the network, and you do not exist to those enveloped in social networks. This directly affects the self's existence.

Most students Freitas interviewed used a variety of social networking platforms, which contributes to the networked and multiplicity of self-presentation. They often did so with specific intentions. For example, students report using Facebook primarily as an overarching profile that will contribute to career and professional goals; they use Snapchat for small-group or one-on-one communication that integrates text, visual, and audio and disappears in twenty-four hours; and others use geolocated, anonymous sites to post about what's happening in their area and often to eschew the need to appear happy because anonymity allows rude, trash-talking, cyberbullying behavior to happen undetected by other users. While cyberbullying and cyberstalking also happen on other platforms, college students are less likely to engage in these behaviors because they have internalized the message that platforms linked to one's real identity directly affect future employment. Some avoid any mention of politics or social action or anything that might reflect or elicit a negative emotion.[37]

These students, like many of us, recognize that we have audiences on social media. In "Branding as Social Discourse: Identity Construction Using Online Social and Professional Networking Sites," Corey Jay Liberman reports on the difference in use between LinkedIn and Facebook. He notes that LinkedIn's "predominant branding method" relies on "posting information about oneself and uploading photos and videos." In comparison, Facebook includes "a quantitative measure of one's friends and the groups to which one belongs, [and] this branding mechanism involves a much more proactive approach to creating and shaping one's socially constructed identity."[38] We cannot necessarily control our audiences in a mediated public space; when this happens, users experience what danah boyd, a principal researcher at Microsoft, has termed "context collapse."[39] For example, a high school student posts a picture on Facebook of themselves at a party, and their parent (who is a Facebook friend of one of their Facebook friends) comments on the picture. Of course, they could have managed the privacy settings of the picture to allow only a limited group to see it, but they forgot. Or a "friend" could have copied the image and reposted it, creating a larger and unknown audience. Similar things happen in person, when, for example, a teacher walks in the room earlier than expected and catches two students imitating the teacher. The intended audience collapsed. The difference is the reach of the in-class experience is severely limited in an unmediated space versus online in a mediated public space. That has significant consequences for impression management and the impact a mistake or success can have on our self-presentation.

The reach of social networks is not always a negative aspect of identify formation. Sara Green-Hamann and John C. Sherblom address how digital technologies often create spaces for oppressed and ostracized groups to be empowered and claim their self-identity, which translates to their flourishing offline. In "Developing a Transgender Identity in a Virtual Community," Green-Hamann and Sherblom write, "social identities are negotiated and developed through the communication processes that occur within a person's social network and community. The communities' values confirm, corroborate, or contradict that person's identity." For individuals who are transgender, offline expression of their identity can be deadly in some cases. Online spaces offer an initial space to explore and live into this self-formation. Their research focused on experiences in the Transgender Resource Center in Second Life, an online virtual community where more than fifteen million registered users interact via avatars in a three-dimensional virtual space using synchronous communication. They found that "as the boundaries between physical lives and social media become increasingly fluid and inseparable, an individual's online identity and community participation interact with and affect the physical one."[40]

The line is increasingly blurred between an online and offline identity. Even when users can manage audiences and information with precision, the medium still affects the process of formation. The experience of anxiety when trying to always look happy and accomplished impacts one's sense of self in ways that cannot be separated between digital and fleshly spaces. Like any new form of communication, impression management on computer-mediated communication has to take into consideration the technological affordances that may be similar to or different from in-person communication techniques that look at facial expression, clothing, location, tone of voice, and use of language and consider new forms of language, such as text messaging, emojis, and use of color, layout, format, video, audio, and filters. While digital communication forms are expanded, they still exist within their own sets of limitations.[41]

Is it morally wrong to portray one's self in various forms? Why, if I have a body with female genitals, must I present as a woman in an online discussion? Should gender matter? If so, who benefits from the enforcement of that self-presentation? This raises significant questions about which identity categories matter when considering the moral formation of self or whether my actions—how I treat others and what attitudes I espouse and support—matter more than my physical form, to which most identity markers are connected. In other words, is multiplicity of identity necessarily antithetical to authenticity or integrity? If we think back to the notion of a trinitarian theology that exemplifies difference in unity, there must be a possibility for authenticity and multiplicity to coexist. Being a networked self is simply a more experiential way to know ourselves as relational. Is it the medium that is perpetuating unhappiness, or is it real and preconceived social expectations generated by economic and political systems? When one gains anonymity online, both experiences of freedom from oppression and increased oppressive behavior are consequences. I am not arguing that these mediums necessarily contribute to moral malformation. However, they do make us increasingly aware that critical engagement with the medium is necessary when considering moral formation of the self.[42]

## THE SELF AS DIGITAL, OR I SHARE, AND THEREFORE I AM

For centuries, Christian theologians have struggled to make sense of human existence as embodied spiritual beings—embodied spirits and inspirited bodies.[43] Now, we must consider how we are digital embodied spirits. That is to say, we are not only inseparable from our online identities, but increasingly, digital technology is an appendage of who we are. Many feel a sense of loss when they cannot find their mobile phone; others attach watches and wristbands that transfer data telling them via notifications what and when to do things as well as keeping communication lines open for texts and calls. The cyborg image that comes to mind most readily is the person who wears their Bluetooth earpiece all day, every day, available for a constant stream of music, news, or phone calls. The meshing of human and machine that digital technology promotes raises important questions for us about moral formation and the virtue of attunement. If our devices or the mechanization of our lives increasingly drives our desires and sense of self, to what must we be attuned and how?

Push technology is the most common way technology grabs our attention and reshapes our behavioral responses. When digital technology began, humans interacted with computers primarily in a pull relationship. We pulled the data we wanted from software or internet searches. Increasingly, software is designed to both pull and push information. Mobile devices push information to the user regularly by posting notifications on the screen of a new email, calendar reminder, or Twitter update. The hardware of phones is also linked to push technology by blinking with differently colored lights to show what notifications are available or making a noise or vibration to get the user's attention. People are not simply obsessed with checking messages; rather, our devices signal to us, maybe even train us, to respond when we see a light or read a partial message. There is a dialogical, or conversational, relationship between the technology and the human that elicits emotional responses such as happiness, laughter, anxiety, frustration, stress, and wonder.

In chapter 1, I discussed the role of algorithms in interpreting the mass amounts of data that are generated by digital technology. In this section, we turn the focus to how that data constitutes who we are as human beings. Roberto Simanowski in *Data Love: The Seduction and Betrayal of Digital Technologies* comments on self-tracking devices and the "smart things" that now guide our living: "Commonly also referred to as the quantified-self, the culture of self-tracking has been developing for years generating products like Fitbit, Digfit [*sic*], Jawbone's wristband, and Nike+, which monitor—and thereby control—the frequency of steps and pulses and thus also how we move, sleep, and eat."[44] The idea sold with these products is the promise that if you can measure it, you can control and change it.[45] Measurable self-assessment becomes the key to a better self, a virtuous self. Of course, that self is defined by market-driven and consumerist notions of health, so attunement is not directed toward self-reflection, or relational growth, or God. With the promise of erasing social inequalities, advocates of digitization promote objective measurements. Simanowski remarks that the quantified self lives by the numbers: "it is only the enumeration of views, shares, and likes that guarantee an equal right to be heard regardless of all differences in education or financial prosperity."[46] This is a democratized equality done by number, based in a consensus view of self-optimization—crowd-regulated, individuality at its best.

The self becomes more and more quantifiable because that is in fact how data is read. "To participate in today's digitally networked world is to produce an impressive amount of data,"[47] and that data is read by algorithms that reduce the user to logics and measurements. John Cheney-Lippold's thesis is the title of his book, *We Are Data: Algorithms and the Making of Our Digital Selves*. He argues that we are no longer intelligible in a digital world as only flesh and spirit. We are "represented and regulated" by data or collections of interpreted data he refers to as measurable types.[48] These measurable types "have their own histories, logics, and rationales. But these histories, logics, and rationales are necessarily different from our own."[49] That is to say, the gender that Google thinks I am has nothing to do with my fleshly body. It is a collation of data points related to my searches, use of Gmail, language content analysis in Google Documents, and user information from an Android phone. If my actions fit the measurable type of male, then that is what I am to Google for marketing category purposes. There is no moral import to gender for Google, only financial (though that has other moral implications). Cheney-Lippold writes, "What algorithmic gender signifies is something largely illegible to us, although it remains increasingly efficacious for those who are using our data to market, surveil, or control us."[50]

The meaning of who we are as digital, embodied spirits is interpreted differently by people in our network than by the companies profiting from our data. In chapter 3, we will discuss how the datafication of the self relates to dataveillance—the use of data for surveillance. Oftentimes, we cannot even see the structures that control us, and overwhelmingly we willingly give our data away. When it comes to data, digital devices are a one-way mirror "[i]n which internet users remain ignorant of how their data is used while site owners are privileged with near-universal access to that data."[51] Take, for example, quizzes users do online that will create a data type of who you are; the input data means something different to the company collecting the information than it does to your Facebook friends, to whom you put out the information and the link to which type of preacher you are, who your spirit animal is, or which celebrity you resemble.

In the process of datafication of the self, or being an embodied digitized spirit, lack of awareness of how social structures such as marketing shape what we see and whom we see online directly affects moral formation of self. In *Everybody Lies: Big Data, New Data, and What the Internet Can Tell Us about Who We Really Are*, economist and data journalist Seth Stephens-Davidowitz has dedicated his work to following the data trails we leave online. He says, "It turns out the trails we leave as we seek knowledge on the Internet are tremendously revealing."[52] For example, he predicted that Trump would win the electoral college votes in the 2016 US presidential election even when sophisticated data-computing firms, such as FiveThirtyEight.com, could not. How? He begins with a simple assumption that everybody lies and we lie the most during person-to-person interactions, such as a polling phone call. However, when typing in a random search on a search engine, we are more likely to be truthful, very truthful. He writes, "in this case, the search window serves as a kind of confessional."[53]

Like most confessionals, we share subtle and explicit details, believing no one will know or share what we have said. Related to the 2016 election, he found a number of indicators related to candidate preference that polls could not calculate. When conducting a search, a voter is more likely to first type the name of the candidate they prefer, as in Trump–Clinton or Clinton–Trump. Across midwestern states, there were many more searches for Trump–Clinton than for Clinton–Trump, predicting that, overwhelmingly, White voter districts voted for Trump. Additionally, people may tell campaign outreach callers they plan to vote, but online searches on "where to vote" or "how to vote" better predict the final percentage of citizens in an area who do vote. For example, voters in predominantly Black voting districts reported they were planning to vote, but there were very few online searches for voter information in these voting districts. Voting in these districts was down in the 2016 election. Rather than simply see this as a fault of geography, Stephens-Davidowitz also found, after four years of collating data on racial differences during the Obama administration, that "areas that supported Trump in the largest numbers were those that made the most Google searches using the word 'nigger.'"[54] He suggests that there is an explicit connection between racism and supportive votes for Trump. These are a number of the factors collated via Google search data that helped Stephens-Davidowitz predict the 2016 presidential Electoral College vote.

Based on his data related to race, he suggests that we need a different explanation for racism in the United States. Black and Brown people can clearly show evidence of incidents of racism. And yet, many White people believe racism does not exist. Researchers have put forward a theory that racism is often caused by an implicit bias—actual biased actions that arise from an unconscious reaction to people or events based on race among other factors such as gender and class. Stephens-Davidowitz shows via his research that "an alternative explanation for the discrimination that African-Americans feel and whites deny: hidden *explicit* racism."[55] Each year, the word *nigger* appears in more than seven million American searches. He controlled for the word "nigga," often used in rap and R&B lyrics. Also, "nigger jokes" are the most searched identity-based joke category. Searches using this term rise when African Americans are in the news, including whenever President Obama gave a nationwide address. He argues, "it's hard to imagine that Americans are Googling the word 'nigger' with the same frequency as 'migraine' and 'economist' without *explicit* racism having a major impact on African Americans."[56] As we consider how the confessional of the Google search box affects our moral formation, I would also argue that Whites are deeply, morally deformed by their own racism.

Overwhelming explicit, active racism is evident when following the digitized aspects of White selves in the United States. Now, those searching for this term may not agree that racism is a deformation of their moral selves. However, that claim is counter to the theology and ethics presented in this book. The theology we have discussed thus far calls for a recognition and celebration of diversity, not a hierarchy or eradication. Also, the joining of one nation to suppress and homogenize all others is what God stands against at Babel (see chapter 1). Christian ethicist Jennifer Harvey in *Dear White Christians: For Those Still Longing for Racial Reconciliation*, describes for the reader how White racial identity is constructed. She says, "Whiteness literally and directly emerged from violence as a socially real, meaningful, and recognizable category"—one created by laws, pseudo-science, economic and social practices ranging from slavery and lynching to policing and incarceration.[57] For White Christians who view violence and coercion as a moral evil, "to be white is to exist in a state of profound moral crisis."[58] White embodiment, even in a digital realm, perpetuates that violence again and again. Harvey argues, "white racial identity has emerged as those deemed white have lived in active or passive complicity with racially unjust practices and have continually accrued, even until today, the material benefits of those histories and their contemporary manifestations."[59]

The reflection in digital device mirrors is a racist America. "Although each of us as individuals interact with technological artifacts countless times every day, the character-shaping properties are especially clear when viewed at the level of community and society," writes Brad Kallenberg in *God and Gadgets: Following Jesus in a Technological Age*.[60] Kallenberg is referring mostly to desire-shaping technologies, such as push notifications, and changes in communication style, but we must also be keenly aware of how digital technologies afford new spaces for our moral deformations to perpetuate and grow. The internet did not become a utopia or bring about the eschatology (God's heaven on earth, the here and not yet). Rather, the inspirited digital body is as morally entangled with sociocultural oppression now as in the analog past.

## ATTUNEMENT AND DIGITAL MORAL FORMATION

Attunement cultivates perceptive attention; adjustment to needs, desires, and feelings of self and other; and sophisticated sensitivities to the world around us. Attunement is an embodied awareness as much as a thoughtful and reflective practice. As we have discussed, in a digital world, one must be informed about how we are digitally constituted in addition to our inspirited bodily existence if we strive for attunement. This requires we move away from an instrumentalist view of technology as a tool we can pick up and put down, entirely under our control; on the other hand, digital technology is not a new paradigm of existence beyond human control or intervention. Kallenberg reminds us, "Technology is neither our dictator (technological determinism) nor merely our tool (cultural determinism) but something much closer to us, under the skin or in the blood as it were. . . . We are social cyborgs. So closely is technology bound to our life together that we must conclude that all technology has moral, political, communal, even human properties."[61]

Attunement responds to the two main qualitative shifts of self-formation in a digital culture—impression management and datafication of the self. Who we are includes the content we produce, but perhaps more important, the *way* in which the content (including our self-formation) is produced. In *The Hidden Power of Electronic Culture: How Media Shapes Faith, the Gospel, and Church*, Shane Hipps unpacks McLuhan's wisdom for Christian ministries. He remarks, related to digital technologies,

> Their power is staggering but remains hidden from view. Because we tend to focus our gaze on their content, the forms of media appear only in our peripheral vision. As a result they exert a subtle yet immense power. By exposing their secrets and powers, we restore our ability to predict and perceive the often unintended consequences of using new media and new methods. This understanding of media is crucial to forming God's people with discernment, authenticity, and faithfulness to the gospel.[62]

Discernment, authenticity, and faithfulness are makers of attunement. However, each is re-envisioned in a digital landscape. Discernment requires a counter-digital-cultural approach to computer-mediated communication, which forms us toward quick, constant, and multifocal response. Researchers suggest that we take digital technology breaks or fasts in the ritually religious sense of the word—intentionally build in time for slow reflection. Others offer simple and probably more realistic suggestions such as turning off notification settings on mobile devices or reducing the number of devices we rely upon that link us to the Internet of Things. As we have discussed, leaving behind or even disentangling the self from the digital is increasingly difficult if not impossible when we zoom out beyond our own use of digital devices and software to the digital social systems that manage communities, nation-states, and global trade for example.

Donna Freitas found two distinct groups in her research cohorts that were not as prone to impression management anxiety on social media. These groups give us clues to the process of attunement. The first group were students at highly prestigious schools.[63] Freitas found students at this type of school had a critical approach to social media not found at other institutions. Some maintained an objective distance, engaging social media only as needed; others discussed aspects of identity creation with literacy of "medium," not just message content. In other words, they assumed a critically engaged and thoughtful stance to social media from their initial engagement with the medium that other students began to consider only once Freitas invited it. Freitas found that few if any institutions integrate social media studies across the curriculum. She notes that students are taught to professionalize their accounts but "are not being challenged to think about social media during their studies" at any of the institutions she visited.[64] Of the students who do critically engage social media, she writes, "these students had come to college with such a high level of intellectual engagement that it seemed natural for them to apply those skills to social media too. It's just what they do."[65] Discernment is not simply achieved by breaks from our digital devices, though that can be a helpful response. We also need to critically reflect on the mediums we use in an ongoing, daily manner.

The second group that Freitas identifies are religiously engaged students. These students' engagement with social media exemplifies the intersection between discernment, authenticity, and faithfulness. "Those students who allow the devotion to their faith to permeate their online worlds use their religious traditions as a framework for navigating their behavior and posts—one they find far more meaningful and sturdier than warnings about future employers and prescriptions for curating one's online image," reports Freitas.[66] These students exhibit a sense of attunement that stems from their process of discerning what God requires of them, how they will engage various social media to present that, and the importance or commitment to this in their lives. Freitas says, "while these students are just as image-conscious and as aware as everyone else that they have an audience, having God and their faith tradition filtering their online decision-making seems to help them stay grounded."[67] These students are not following a preset list of rules given to them by their churches. Rather, they are able to critically engage their digital self-formation because they have a "sturdier" moral sensibility about the world around them, their place in it, and how this relates to God. The religiously affiliated students' version of authenticity does not mean they have only one way to present themselves. In fact, they sometimes use a variety of group and privacy settings to distinguish who sees what post. They have figured out that a networked self requires sophisticated sensibilities to build and maintain relationships that will nurture God's presence.

Social media is not the only place we are formed as embodied digital spirits. However, these two examples give us clues to how we must attune our moral selves. We need to assume a stance of critical engagement with everyday digital interactions that questions the medium. Second, we need to ground ourselves in a moral community and relationship with God that informs our sense of self and formation of relationships regardless of platform or space.

Some theologians argue the latter is not possible in a digital world. Christian theologian Quentin Schultze writes about the need for "shared commitments to truthfulness, empathy, and integrity" in order to "regain authenticity in an information society."[68] The lack of physical identity markers or authenticators in cyberspace led Schultz to claim, "It is increasingly difficult in cyberspace to know who says what, what he or she really means, and whose self-interest is shaping online rhetoric."[69] Schultze is writing in the early 2000s prior to an explosion of audio and visual technology that is the foundation for most social networking. As well, research finds that more often than not, users represent themselves in a manner that coheres with their offline selves, even if they err on the side of more favorable self-presentations (as we do in initial meetings in person). Virtue is often characterized as doing what's right even when no one is watching. The opposite may be true in a world of social media, where it seems everyone is or could be watching even when we think we have limited our audience. Thus, self-narratives tend not to produce a chasm between online and offline representations. However, social phenomenon such as fake news should still lead us to question the authenticity of third-party information. This is different than self-presentation.

For other theologians, digital technology raises a more significant issue related to embodiment and our ability to ground ourselves in a moral community and relationship with God in a space that is not enfleshed. Early Christians railed against the constraints of the body. Some even denied the body as part of the baser, earthly existence from which they sought freedom. Christian beliefs about the body and ethics in response to our embodiment have varied across history. Incarnational theology—Jesus as God coming to humanity in embodied form—affirms the body as a morally significant and necessary part of our creat-edness. So rather than shunning the body or creating a list of behaviors to control its desires, we seek embodied, emotional attunement with ourselves, others, and God, which deepens the relationship between body and spirit while balancing commitments in relationships.

Much anthropological investigation in the early 2000s considered cyberspace a place that lived in the now of instantaneous communica-tion and a bodiless, nonphysical space. As digital technology has grown, we see more and more reliance on digital expressions of embodiment, such as video and audio, not to mention that there is an everlasting memory of each data point that represents our digital life constituting a new form or body.

We heard from Kallenberg at the start of this section claiming that we are social cyborgs and technology is under our skin and in our blood. In chapter 2, titled "Is Technology Good News?", he considers the impact of communication on evangelism and whether digital communi-cation can be as effective as in-person communication. He suggests that it could but rarely is. In chapter 3, "The Technological Evangelist," he argues that the Christian minister and ministries need to incarnate the word as Jesus did, which is labor and time intensive. He claims we need bodies and in-person conversation to incarnate the word so it does not become Gnostic and unidirectional.[70]

Christian educator Gordon Mikoski, in an editorial for *Theology To-day*, agrees with Kallenberg.[71] He investigates the practice of commun-ion-on-demand websites, where communion can be experienced out-side the physical church. After touting the benefits of various online worship practices, such as live streaming worship or using audiovisual software for pastoral care, he draws the line at online communion. Eucharist, he believes, requires our real, bodily presence: "The Euchar-ist never takes place in the abstract. It can only be celebrated in particu-lar places and times."[72] Mikoski connects his conclusion to the incarna-tion, writing, "if disembodied communication were a sufficient strategy for relating to the human family, there likely never would have been an incarnation. God could have continued to communicate with the hu-man family virtually and remotely through a series of voice messages and disembodied holographic images (think burning bushes and still, small voices)."[73]

Mikoski's points are persuasive and lean toward earlier claims I have made about the "scandalously particular, embodied, and contextual" nature of God incarnate in Jesus.[74] It is at exactly this point at which the digital self presents new theological insights. Our systems and ourselves are increasingly digital, which does not make us any less embodied or spiritual. Only a few people lived alongside the incarnate Christ as an embodied human. And yet, in our time period, our experience of the incarnation is no less real, embodied or communal, because we experience it through relationship with those around us and the narrative we have in scripture. As networked selves, we can view a trinitarian theology with greater creative possibilities than in the past. The incarnation is central to one aspect of how God lives difference in unity. We ought not to forgo the aspects of our religious practice that exemplify that embodiment. Yet God is also the burning bush, still small voice, tongue of fire, breath over the waters, and many more nonfleshly embodiments we do not, perhaps cannot, yet perceive. We will explore this in chapter 4 as we turn to questions of digital technology, ecology, and environmental degradation.

God's presence is communication in multiple forms. Theologians have often played off the presence of the word *communion* in communication as part of an understanding of God. Talking specifically about humans and our relationships, Kerri Harvey in *Eden Online* writes, "Communication then, in both its process and its products, is the canvas on which the converging forces of time, space, and culture paint various constructed and socially functional portraits of the other and of this self."[75] I would add, of and with God. Communication takes time, location, and physicality. The experience of these three is altered from analog to digital, but it is not lost.

Why does attunement as the process of moral formation make more sense when discussing our digital lives? Digital existence heightens our connectedness and relationality. Attunement locates virtue in relationship and requires an ethical approach that confronts personal "narrative diversity and the otherness of others."[76] Digital interaction and integration are changing so rapidly that most of our "sets of rules" are outdated within months of developing them. Additionally, digital technologies bring us into contact with various cultures and geographic locations, at times creating new ones. The digital expands a sense of self beyond the outer casing of our skin and into worlds and lives we experience in new ways. Attunement requires awareness and attention to internal and external dynamics in an ongoing, critically discerning manner rather than a static or predetermined formula.

Digital is not a new utopia. We must recognize, understand, and liberate ourselves and society from racism and other moral deformations of our digitally embodied spirits if we are to live into God's example of difference in unity as the *imago dei* of a networked self. Attunement requires emotional, embodied, and spiritual awareness as much as responsive digital affectability. As a historically rooted and diverse community, Christians bring a wealth of moral insight to this process. In some cases, we must shed old forms of moral practice that no longer produce attunement in today's digital world. In other cases, like the one addressed in chapter 3, moral practices of metanoia are fertile ground for attunement in response to a networked self in a digital landscape that never forgets.

## EXCURSUS 2: MORAL COMPLICITY IN THE DIGITAL SOCIETY

> And they clothed him in a purple cloak; and after twisting some thorns into a crown, they put it on him. And they began saluting him, "Hail, King of the Jews!" They struck his head with a reed, spat upon him, and knelt down in homage to him. After mocking him, they stripped him of the purple cloak and put his own clothes on him. Then they led him out to crucify him.
>
> —Mark 15:17–20, NRSV

I follow Nigel Hayes on Twitter. He is a star basketball player at the University of Wisconsin at Madison, my alma mater. In September 2016, he posted the following, "Racism toward black people isn't getting 'worse,' it's getting filmed and shared for all to see what actually goes on. #BlackLivesMatter."[77] His tweet was sparked by social media sharing of shootings of Black and Brown men, women, and children by police in the United States. Hayes's comment can be read as both a reality check for the White viewer for whom the video is needed to affirm that such violence is real and for Black communities for whom it may retraumatize, silence, and enrage.[78] In this context, what does solidarity look like for me as a White woman or White people more generally?

Solidarity is more than recognizing the suffering of the other in their humanity; we have to do something in response. Participation in social networks makes it possible to share and consume images from racially comfortable locations, similar to how Christians might view the cross today apart from the real, intense bodily suffering Jesus endured. Moral theologian M. Shawn Copeland might call this "the new imperial deployment and debasement of bodies."[79] In her book *Enfleshing Freedom*, she writes, "Through a praxis of solidarity, we not only apprehend and are moved by the suffering of the other, we confront and address its oppressive cause and shoulder the other's suffering."[80] Jesus, in flesh and blood, is not only moved by the suffering of the other but also takes on the suffering of the other.

How might I cultivate responses that deepen solidarity rather than dismiss the reality being mirrored back via my mobile screen? My social networks make public my self-concept; they identify part of who I am. As chapter 2 suggests, we are not separate from our digital presence. As a White woman, if I repost a video of police shooting a Black or Brown person, does my post show that I'm a racially sensitive and aware White person in solidarity with Black communities? Does the intention behind the post ethically outweigh the consumption built into the digital system? What about the reduction of a person to a hashtag?

Rather than repost racial violence and suffering, I am challenged to affirm this reality without needing video footage, admit my role in it, and post about ways I will actively resist perpetuation of and participation in racial violence and suffering. This type of response leads to attunement—a process of moral growth that is responsive, accountable, and expansive—discussed in chapter 2. What will you do next time such a story or video is in your social media feed?

# 3

# MORAL FUNCTIONS BEYOND THE DELETE KEY

The digital data we produce increasingly shapes who and how we are in the world. Whether participation is by choice in social networks or by requirement in data systems of government, schools or health care programs, who we are is transformed into data, interpreted and reinterpreted. This mass amount of data is cataloged and stored with every click and share. It is also available to an ever-widening public, including companies but also networks of people. The mediated nature of digital technology transforms divisions between public and private information, raising questions about surveillance and privacy. As we consider the moral formation of individuals and communities, we must ask about the impact of living in a world where our pasts are cataloged and remain forever available for interpretation.[1]

We are aware, in this new digital culture, every share, post, or comment is archived, creating an online trove of information that identifies each person (i.e., where we get our news, what we look like, what sports team or music group we like, where we go to church, etc.) and labels communities—who visits and "likes" the church website, who has a "support X cause" Facebook photo overlay, and so on. This digital trail was not entirely expected. In the early days of the World Wide Web, "Internet utopians claimed the Internet would give everyone the power to surveil, to see and not be seen, to become a bodiless and thus unseeable user," writes digital media scholar Lisa Nakamura. "Instead, we have become more visible and trackable than ever."[2] And as Nakamura points out, we've come to depend on the hypervisibility of ourselves and others "for access to friends and community." There are significant and positive outcomes of hypervisibility, such as grandparents meeting a new baby via video call or parents staying connected with children who move away, an opportunity for a young transgender teen to find a supportive space while living in an isolated area or young adult advocates for Black Lives Matter organizing nonviolent protests.

Of course, that is the information we actively *share*; by using platforms and services such as smartphones, iTunes, credit cards, Instagram, search engines, Snapchat, and so on, we are also agreeing to let these companies collect data from us, a process known as dataveillance. Our clickstream (opened programs and links), friends and followers lists, shared information and images, location, and personal demographics used to establish a profile comprise the majority of the data collected. And this data becomes a valuable product. Platforms sell information data points to advertising companies who can tailor a product to your interests or generate the kind of content that keeps you returning. This is partly why social media platforms are designed to increase and invite sharing: the more we share, the more data is generated and the more profit is made. In other words, the advance of digital culture simultaneously launched an economic system in which businesses seek profit and we seek personal and community connections.

Digital literacies related to dataveillance and social surveillance provide a robust understanding of our digital trail and its impact on ethical living. Digital trails raise questions about how we live in a world that increasingly does not forget. More and more frequently, we are asked to wrestle with issues of data archiving and dataveillance that comprise our digital trails. Some argue for a renewed practice of forgetting both from a legal perspective of erasing data and from a social and developmental perspective in terms of enabling users to change over time. How as Christians do we understand the call to change on personal, social, and ideological levels as well as the role of forgiveness in a world where forgetting has become a luxury?

Moral formation is a process of change, growth, and development. When we think of our digital lives, we might envision them as dynamic and progressive, constantly changing at a rapid pace. However, our digital lives are much more akin to a process of accumulation subject to external forces, such as algorithms, predictive analytics, and surveillance systems that never forget. Some might say this makes change difficult because we cannot shed the past and move into new futures. Moral formation is a process of educating and recommitting ourselves to renewed morally good ways of being and behaving in the world. We do accumulate a narrative of our moral life. The purpose of that narrative fuels our accountability to continued change, rather than tethering us to an immutable past.

The cataloging of digital lives does not have to work against the Christian call to transformation and change—metanoia. In the Christian tradition, metanoia describes the process by which faith positively enables the human capacity to make change. Part of that process is an accounting of past actions, not a forgetting of them. Digital technologies thrive off of data that relates to almost every aspect of our lives. That data both generates a massive archive and allows for various types of surveillance. I will explore the ways that dataveillance and social surveillance shape digital interactions and possibilities for change. Many who remark about the endless memory of the internet are concerned that keeping people's histories will lead to a notion of stagnation and accumulation in the shaping of moral character. Returning to metanoia, I detail how it has to be lived out in a responsive manner and evidenced in reformed actions. Metanoia with its insistence for accountability to the past addresses both individual change and recognition of structural and systemic injustices that must be disrupted through daily practices.

## SIN AND METANOIA IN A DIGITAL AGE

The theological and ethical concept of metanoia—marked by change in one's life that reflects a deep penitence or conversion—is often shorthand for a "turning away from sin and toward God." As Christians, we are regularly striving for metanoia. When we reflect on our sinfulness, we often ask forgiveness from God and/or afflicted parties whom we have harmed. This happens many times in our lives as Christians because we are imperfect. The accumulation of these "turnings" accounts for much of the developmental story of moral formation. With each responsive moral act of forgiveness and accountability, we cultivate selves less likely to repeat the harm again. This process is not done in isolation. Relationship with God and communities of people who surround us, such as family, friends, and our faith community, significantly impact metanoia. Change requires more than a thought or feeling to be altered; concrete behavioral action is necessary, and thus each act of changed behavior renews one's commitment.

Metanoia is as much about individual responsiveness to God as it is about relationships with others. The recognition of social and relational influence opens up a notion of sin from personal or individual alienation from God to sin as alienation from one another. The relational aspect of metanoia suggests an understanding of sin that is deeply shaped by social context. It also calls attention to the historical and collective construction of sin that often seeks to hide social or systemic forms of sin.

For example, theologians have for the past century debated the gendered nature of sin. Pride, which is what some might associate with popularity on social media, has been discussed as a male vice—an overestimation of one's importance. Whereas self-sacrifice to the point of erasure of self, like becoming only what others want to see rather than your true self on social media, has been considered women's sin. These distinctions in how interpersonal sin manifests itself are helpful; however, they also reinforce systemically oppressive notions of gender that naturalize distorted relationships. Men who do not exhibit pride in the form of excessive self-confidence are socially penalized as are women who do not submit to male authority and accept passivity. The reinforcement of gendered notions of sin rather than recognition of them as historical constructions of social relations is one example of the connection between interpersonal, social–historical and ideological–cultural dimensions of sin.

My understanding of sin and metanoia differs from predominant historical approaches. Many theologians focus primarily on sin as personal failing and alienation from God. Humans are set on this path of alienation beginning with the fall of Adam and Eve, and it is then biologically inherited by birth (original sin), further gendering notions of sin. Through the sacrifice of Christ on the cross, we participate in a reconciliation with God that opens humanity to the possibility of repairing our alienated state. Thus, metanoia would be a turning toward God and away from one's own sinful nature. This classical view of metanoia offers very little attention to corporate responsibility or systemic analysis. It leaves us in a personalized cycle without much accounting for systemic influences on moral choices or behaviors.

Helpful to an understanding of how sin, or distorted relationship, functions in interpersonal, social–historical and ideological–cultural dimensions is the work of feminist theologian Rosemary Radford Ruether. Ruether moves away from a focus on sin as a biologically transmitted brokenness to an "inherited, collective, historical dimension of sin."[3] Through this understanding, she defines metanoia as a "process of emancipation" from oppressive social and cultural systems "to create a new self and a new society."[4] In Ruether's view of sin, there is still individual human freedom and accountability, but she recognizes that moral choices do not exist in a personal vacuum. It is widely accepted that social, cultural, and religious systems shape the personal and the relational. Thus, Ruether says, metanoia requires "we not only have to confront our own sadistic and masochistic tendencies, but also have to unmask the claims of the dominant culture that misleads us about the nature of good and evil."[5]

The nature of good and evil is extremely difficult to discern. Some might argue that humans are incapable of this moral task. On the other hand, Ruether suggests human capacities are "ambivalent rather than depraved or in an irreparable condition of alienation." Ambivalence means as humans we are pulled equally and strongly in opposite directions to be and do good as well as to be and do evil. The tendency to do evil is shaped historically by systems of evil in which we sometimes unintentionally and often uncritically participate as we choose negative, life-denying directions.[6] Violence and evil are sustained by individual actions that compound and reinforce patterns and systems that perpetuate interpersonal as well as social alienation. At times, it can seem as though we cannot break this cycle. Affirming that there is human freedom to choose to enhance life (good) or to stifle life (evil) calls us to accountability. We have a choice, a moral choice. God's grace can help us with that choice, leading us in positive, life-giving directions. We are still responsible, and human agency is required. It is the tendency to do good that connects us to a more authentic and true nature, "our 'imago dei.'"[7] Either choice, to do good or evil, does not exist in a personal vacuum.

Sin is using another person or creation in a manner that denies the interdependency and mutuality of creation. Such instrumentalism can be directed toward oneself or others. Consider the potential reduction of relationship on social media to numbers of likes or follows. Yes, people have reduced their relationships to a numerical value—that is a personal choice to turn away from the robust quality of relational intimacy. Yet that is not the full story. "Relationship" is redefined in the digital environment by systems of capitalism that leverage the desire to be liked for profit. Awareness of how digital technology reshapes this desire and the economic motives behind it is necessary when considering how to ethically evaluate this sin or alienation from others. This is a way digital technology reshapes our patterns of sin. In what other ways does it affect acts of forgiveness and movement toward metanoia in the new social landscape of digital technology? First, we need to better understand the impact of endless memory and never forgetting as well as the surveillance practices afforded by digital technology.

## DIGITAL DATA, ARCHIVING, AND SURVEILLANCE

Sharing and archiving of information are shaped in new ways by digital technology. Social networking is a term coined in the internet era but is truly a historical practice.[8] Forever, humans have shared information, including gossip. The digital aspect of social media and networking, however, breaks down the geographic, time, and archival aspects of our historical patterns of communication. Social media platforms are instead designed for openness, collaboration, and networking. Take journalism, for example; not more than twenty-five years ago, most journalists' work was printed in magazines and newspapers or broadcast on television. The information was pushed to audiences who passively received the information. The only space for interaction or reaction was writing a letter to the editor in hopes of publication days or weeks later. Now, journalism is an open field, often being produced on the ground in real time as participants record and report direct experiences. Anyone on these platforms can respond and share the information instantaneously. What makes social media different than past participatory practices is the extent to which it "allows audiences to talk among themselves, to critique, remix, and redistribute content on an unprecedented scale."[9]

## Digitally Mediated Communication

Mediated communication has taken place for years—writing a letter requires paper and a pen, which mediates the communication. Shifting the mediated form of that letter to email exponentially expands the reach of that communication, including into public spaces. Unmediated communication is face-to-face, in-person communication, though some might argue that clothing choices and gestures mediate in-person communication. For the purposes of this discussion, mediated communication refers to information mediated by digital processes that shape communication. Digital processes have altered time and geographic constraints that once existed in an unmediated society. That is to say, prior to digital technology, communication was restricted in very particular, physical ways. Even though the writing of a letter is mediated, it does not bring about a mediated society or mediated publics because the letter remains restricted by its materiality. The Gutenberg era, a period of human communication marked by a dependency on print, authorship, linearity, fixity, and closure, used forms of mediated communication, but the social context was still mostly unmediated. Digital technology launches us into an era of communication and information that brings about mediated publics marked by openness, collaboration, and easy access to archives.[10]

In the twenty-first century, we live in *mediated publics* regardless of our personal choices about digital devices or social media use. Digital social media takes down the walls of public spaces that once contained our experiences, sharing, and storytelling. Social media researcher danah boyd suggests that mediated publics are created by *four affordances*—aspects that encourage certain practices—which make social media unique. Affordances help us see what can be leveraged or resisted in a social space. Specific to social media, boyd addresses changes to persistence, visibility, spreadability, and searchability of information.[11] Persistence is how long information is available, visibility is how widely information is shared, spreadability is how quickly and easily it can be shared, and searchability is how information is archived and made accessible. These possibilities or affordances are what constitute a mediated public or publics in which we now live. These four aspects of digital technology are the basic components of what we now call *going viral*. When the information is helpful and positive, the ability to go viral is welcomed. On the other hand, when the information is derogatory or condemning, going viral only exacerbates negative effects.

Consider photographs for a moment. When Jenny was in high school in 1990, pictures were taken with a camera, negatives were developed, a store printed the pictures, and she got as many paper copies as she could afford. If Jenny were lucky, only 15 to 20 percent of the photos had to be thrown away because they were out of focus, someone's eyes were closed, and so on. There was no checking to see the quality of the photo before development. If someone had the negative of Jenny's photo, they could reproduce it through this same process. Yet that was limited by physical copy production. This high school example is an experience of an unmediated public, which is different than saying there was no mediated communication. There was, but the "public" related to a photo was unmediated. The public related to Jenny's photograph—direct participants and those who see the picture afterward—is controlled and limited. After Jenny takes the photograph, she has to intentionally give the photo out to specific people to generate a public audience. Jenny cannot completely control who will see the photo, but the visibility and spreadability are low. Eventually, the photo will no longer persist as the paper wears out and the negative is lost or shrivels, thus reducing the persistence of this image. And there is no searchability related to Jenny's photo, other than perhaps through a personal photo album kept on a shelf inside her home.

Almost three decades later, we are living in a digital age. Now, one of Jenny's high school friends scans a physical copy of that photo, creating a digital file. The digital copy of the same picture has no physical limitations. The friend posts it on Instagram and tags Jenny in the photo, resulting in the picture showing up on her Instagram feed. Jenny untags herself because she does not want that photo "public" for her friends to see. Unfortunately, that only prevents her current followers from seeing the photo. Many other networks may see the photo and be able to copy and repost it. Jenny can control only whether she is explicitly identified in the photo on this platform. In this process, Instagram keeps a backup copy of the photo file as does the storage system and mobile service provider of the friend's phone. Once the picture becomes a digital file, it becomes mediated (as in, turned into a different form of information because of its ability to be shared, reproduced, and spread in ways a hardcopy photo cannot). Thus it has the potential to expand into various publics. The photo will now persist as a digital copy in multiple locations even if it is not widely shared or accessed. The photo has the potential to spread and be visible as old friends share the photo with new networks, leading to a very different existence in the mediated public of digital culture.

In some cases, one photo or comment seen by the right (or wrong) person, such as a boss, a romantic partner, a media outlet, or an internet troll, can have significant ramifications. Fortunately, in the case of Jenny's high school photo, she probably does not need to worry about it going viral, whereas most public figures do. We often assume the internet is a democratized environment where we all participate equally given the open access to most platforms from mobile devices and even public computers on public wireless networks. However, this is rarely the case because users have particular opportunities to leverage the affordances in different ways, as do businesses.

Regardless of whether content goes viral, each of the shares, likes, and reposts creates a digital trail. Awareness of the impact of affordances such as persistence, visibility, accessibility, and spreadability via the internet and especially social networks shows the digital shift of information and communication from largely an unmediated public to mediated publics, in which we now reside. Awareness and adeptness in responding to the impact of digitally mediated publics increase digital literacy as we strive for liberative ethical practices in a digital society. In particular, massive amounts of data are archived to maximize these affordances as a foundational component of how the internet functions. Data archiving is directly related to surveillance practices that benefit commercial interests and interpersonal social connection.

## Data Archiving

The internet is a massive digital archive built through sharing of information. The digital culture we have created raises all kinds of ethical questions about how we protect data, especially when it is the primary way we are made intelligible in a computational world. In a recent Pew Research Center study, most adults in the United States say that "being in control of who can get information" and "what information is collected about them is important" (93 and 90 percent, respectively). But "When asked about search engine providers, online video sites, social media sites and online advertisers, the majority felt 'not too confident' or 'not at all confident' that these entities could protect their data." And less than 10 percent had made any changes to their online behavior.[12] We are simultaneously concerned about the use of our data and confused or cynical about our ability to respond. As we explored in chapter 1, I ask again, are the benefits of digital technology worth the various costs of willingly giving away data?

When we look closely at digital data trails, we can identify at least two distinct practices of surveillance—that which is conducted by institutions and that which comes from our peers. Both sources influence our behavior and choices, which must be accounted for as we expand an understanding of moral choice to the impacts of social and historical context, including mediated publics. The basic structure of digital media is built on the notion of archiving and accessibility to archived information. While digital information can be hidden or settings can be used to protect data, the default foundation of digital media is to be open and accessible. Whether we use the internet to promote our work or engage in private social networking sites, we are exposing ourselves (as data) to surveillance. Data archiving and surveillance are key aspects of the endless memory of the internet.

## Dataveillance

Because of the monetary pressures on social media companies, sharing and collaboration can quickly become cooptation as the process of gathering data turns the user into product. Social media exists in a capitalist market by collecting the data trail of every user and selling that information for compensation. For example, a user checks in on Twitter at a city park using the connected GPS locator on their phone. This information can be used directly by either Twitter or the user's service provider to let advertising companies know this user likes outdoor spaces, specifically in this city. The user may begin to see more advertisements related to outdoor activities or receive information about sales at restaurants or boutiques close to the park. Such data collection is a form of surveillance. The term *dataveillance* refers to the monitoring of a person's activities through digital trails or markers (such as mobile phone GPS, credit card transactions, email, and social media use). The longitudinal accumulation of these trails, or tattoos,[13] is a lucrative resource. Data is turned into product to be sold and utilized to grow business. This is partly why social media platforms are designed to increase and invite sharing. The more we share, the more data is generated and the more profit is available.

Many users argue that this is a small "price" for not having to pay actual money for use of most social media applications. In *Big Data: A Revolution That Will Transform How We Live, Work, and Think*, Viktor Mayer-Schönberger and Kenneth Cukier detail how "data became the raw material of business, a vital economic input, used to create a new form of economic value."[14] They are careful to note how data is often the impetus for innovation and development of new services. They also raise significant concerns about the growth of stored information and the processing power of digital machines. The more we use, the more data we generate—free labor for a growing consumer business model. The collation of data through the collaborative uses of social media in particular raises concerns not only about who profits from the data but also how as communities we will be accountable to one another in a social context ruled by data that never forgets and the tendency to judge people based on propensities rather than actual behavior.

The social-shaping effect of digital trails can have a cumulative effect on users that leads them to give away the role of discernment in decision making to predictive analytics. For example, when we go to a movie-streaming service such as Netflix, we are given movie recommendations. The company uses personal data related to the login, like age, geographic location, household income, and past movie selections. With this information, the streaming service predicts other films the user would enjoy watching. This is a useful service because searching through four thousand movies is inefficient. However, the predictive data used to determine the suggestions often shapes desires in ways of which we are unaware. We may begin to see ourselves as the comedy type only and not explore a documentary or drama because we have success and investment in the accuracy of the suggestions. Of course, related to movies, predictive analytics and social shaping seem more efficient than harmful. What happens when similar practices are applied to medical decisions and linked to insurance costs? The ability to change behavior or escape past data becomes a more serious concern. As users, we may assume that we completely control the experience of whom we friend, how we search, or even what we buy. However, data collation and predictive analytics are in fact influencing us by assuming our desires and thus driving us to content that shapes our digital identity and in turn influences our offline identity.

## Social Surveillance

Dataveillance is not the only form of surveillance that shapes users. While many of us may only occasionally recognize the impact of dataveillance, we are probably very aware of social surveillance. Communications and social media researcher Alice Marwick describes social surveillance as "the ongoing eavesdropping, investigation, gossip and inquiry that constitutes information gathering by people about their peers, made salient by the social digitization normalized by social media."[15] The participatory nature of social media means that users are both watching others and aware they are being watched. Social surveillance rests on a notion of power flow across relationships, peer-to-peer structures (not institution-to-peer, which relates to dataveillance), and reciprocal watching. This surveillance structure can lead to a false sense of equality among users; "clearly not all social media users *are* equal, and the 'democratizing' possibilities of social media are severely limited by economics, citizenship, gender, censorship, and the same processes that limited participation *off*line."[16] As users experience social surveillance, they can become more aware of the inequalities preexistent in their networks.

Minorities, whether based on gender, race, or ethnicity, disproportionately experience online harassment.[17] Well-known examples of harassment include rape and death threats as well as publication of personal data, such as home address and phone number, which targets vocal feminists who speak out against sexualized violence and misogyny in video gaming. Anita Sarkeesian is perhaps the most famous example of this abuse.[18] In 2016, Leslie Jones, an African American actor, announced she was leaving Twitter due to the racist and misogynistic tweets she had received. On Twitter, she wrote, "Ok I have been called Apes," "even got a pic with semen on my face. I'm tryin to figure out what human means. I'm out."[19] National data shows that females face increased rates of online harassment from complete strangers often working in organized groups as well as known acquaintances in the form of stalking and domestic violence.[20] Granted, not all forms of social surveillance lead to such violent or hateful ends. Yet social media creates a space where significant harm can be caused to individuals and communities when hateful and oppressive rhetoric threatens participation and safety. Social media is not unique in this regard, though the effects are often exacerbated by the viral quality of information. Those who defend such actions in the name of free speech do not have a sense of the violence that verbal and visual images have as a form of psychological abuse and intimidation that limits physical movement. Or perhaps they do, and that is precisely the purpose of defending such attacks.

More benign forms of social surveillance result in popularity contests, which can also change our self-perception. Research has shown that social media users edit their information and craft identities. This crafting, while a part of everyday interaction online, sticks in a way that offline behavior does not. This sticky content reshapes memories that might have been "airbrushed" to seem happier.[21] Self-editing may be a process that often bends toward projecting a more positive image in light of cultural norms of popularity, status, and success. In a study conducted by Alice Marwick, she explains how users surveil each other to boost their own popularity, which then confines them to particular self-presentations. She writes,

> Social surveillance is reciprocal. People create content with the expectation that other people will view it, whether that means editing their own self-presentation to appeal to an audience, or doing something controversial to gain attention. But in order to best be seen, the users in this section must monitor the information of others to create the right context. By understanding her 6,000 Twitter followers, McCarthy can decide whether or not they want to know about her "drunk friend in her bathtub." The more that Jackson learns about his audience, the more likely he is to achieve the visibility he craves. Matthew's classmates stalk their peers in order to boost their own social status. The reciprocity of social surveillance engenders both disclosure and concealment.[22]

Social surveillance explains the need to sustain numbers and grow one's following, resulting in a goal of "measurement over meaning."[23] Of course, in other circumstances, social surveillance reflects sharing that provides support and outreach for those who experience isolation due to a number of social factors. Marwick notes that in those cases, social surveillance increases digital intimacy and builds social capital by strengthening networks and relational ties.

The cumulative effect of digitally mediated communication on everyday living is massive. Mediated publics exist because of the affordances of digital data. The effects of information and communication unbound by the boundaries of time and space lead to immense opportunities for connection, surveillance, and profit. The digital data trails produced by our connections and surveillance of each other are utilized in a reciprocal manner as dataveillance impacts user choice. As social media use becomes pervasive and digital media collates the data of everyday living, how are we to responsibly respond? If the internet does not forget and digital trails follow each of us, what is required of those we meet and with whom we are in relationship?

## CTRL+Z: TO FORGET OR TO FORGIVE?

In response to the ubiquity of the internet and its incomparable ability to remember, a European Union court handed down the "Right to Be Forgotten" ruling in 2014.[24] In an effort to honor that ruling, I will not repeat the personal information associated with the case. Simply put, the European Union court ruled in favor of a man who requested that Google remove references to a financial issue he had had in his past. Information revealed in internet searches continued to plague him even after he had rectified his financial situation. While countries in Europe have stricter privacy laws than most countries, the ruling sought to weigh the right of freedom of speech specifically related to media or press freedom against the harms that archived data might present to an individual. The need for such a law and now a protocol to have information removed from the internet is new in this age of digital data. In the not too distant past, finding a news article about someone required a trek to a library, physically leafing through archived papers and pictures, and reading microfiche in the hopes of finding something. Now, search engines such as Google, with simple clues such as name, subject matter, date, or geographic location, return thousands of results in seconds.

The ease of access to digital data about an individual, community, restaurant, movie, school, or almost anything appears at face value to be an unfettered benefit to society. Peer-to-peer use of this information to rank and classify things is social surveillance, as described in the previous section of this chapter. However, when information is negative or harmful, it has exponentially greater consequences due to the indelible nature of digital data. High-profile examples of the impact of digital data sharing and access often result in loss of career or social status. This phenomenon is becoming more and more common. *Rolling Stone* ran a feature article on "A History of People Getting Fired for Social Media Stupidity," which covered relatively unknown folks to celebrities such as Paula Dean and GOP staffer Elizabeth Lauten.[25] These internet lists are often meant to be funny.

On the contrary, they are siren calls to readers about what could happen to them, and they serve as social commentary on the state of digital responsibility. The scenarios detail crossing professional boundaries by posting private client information and activities; destroying property; gloating about violent, racist, or sexist behavior; and mocking employees or customers. Issues range from using foul language to critiquing company uniforms to posting photos of dying patients without permission. A single tweet, photo, or comment is grounds for suspension and often for being fired. Lists such as these abound on the internet. Any one of the featured individuals will live forever with their name and story publicly attached to such incidents. This raises critical questions about informational privacy and public accountability. When social surveillance reveals immoral behaviors, we must consider whether erasing the information is the most appropriate response.

Two well-known examples of viral media reputational demise are Anthony Weiner and Ray Rice. In 2011 and 2013, politician Anthony Weiner's career was cut short as sexual images he sent to women via social media surfaced. In 2011, Weiner, then a New York Democratic representative to Congress, sent a sexually explicit photograph to a woman via direct message on Twitter. The image made its way to a larger public audience, and Weiner eventually resigned from his office.[26] Then, as he was making a political return in the New York mayoral race in 2013, more photos surfaced that he had sent under an alias. His political career has not recovered.[27] Cases of sexually explicit material or even extramarital behavior are not new to politics. However, in the age of digital media, it is much easier to send photos, find photos, and share them widely. Hiding such behavior is almost impossible, even with an alias. In fact, a simple search of Weiner's name produces multiple stories related to these incidents. When searching his name, these are the first references that appear, demonstrating that these events still define "who Anthony Weiner is" for the greater public.

The publishing of a photo or comment is not always by the author. There is now a ubiquity of cameras in buildings for security and in people's hands for leisure. On an evening in 2014, Ray Rice, a Baltimore Ravens NFL player, struck his fiancée (now wife), knocking her unconscious in a hotel elevator after an altercation between them. Rice was arrested and charged in criminal court as well as given a suspension by the NFL.[28] Later, the media outlet TMZ (Thirty Mile Zone) recovered the video surveillance from the hotel and aired it again, and again across media stations.[29] Other stations picked up the released video and aired it as well. With more than 11.5 million views on YouTube, the incident lives on. The public reaction to the footage led the Baltimore Ravens to release Rice, and the NFL suspended him indefinitely.[30] He has yet to make a return to the NFL.

The various ways affordances such as persistence, visibility, spreadability, and searchability are leveraged are why in 2013 Weiner had to answer for pictures sent via social media in 2012 and in part why Rice has not been re-signed in the NFL. For example, when TMZ acquired the video footage of Rice, it already had a significantly larger network than most individuals and companies. It has employees tasked with making sure its information is spread, shared, and visible in multiple formats. As a "journalistic" platform, its content carries a particular authority that a personal posting may not. Thus, TMZ has capacity, labor, and authority with which to leverage the affordances of social media that an average user does not. This allows them to maximize the digitally mediated public, creating viral visibility. The same is true about Weiner, though Big Government News and The Dirty do not have as wide an audience as TMZ.

The above incidents show the amplifying effects of social surveillance on social media. Weiner and Rice transgressed boundaries and violated ethical expectations of their professional role and their personal commitments. The viral effect of the mediated public and active social surveillance in which their information was shared has in certain ways held them accountable for their transgressions. They are accountable to those they personally harmed, their partners, as well as to a wider public. Some might say they should be accountable to a wider public because in some sense they are public figures and thus their behavior has a greater social impact. This points toward the historical and social nature of sin. I would not say they are more accountable because they are public figures. Rather, their transgressions were not only against their partners. For example, Weiner not only violated his wife's trust but that of his constituents and his political office. A similar argument could be made related to Rice and his role-model status as an NFL player. Either way, the social and historical character of sin is more visible in their cases because we acknowledge the connection between personal choice and social systems. But this is true of all sin, not just transgressions for public figures.

Should the indelible nature of digital technology result in Rice or Weiner living lives forever defined by these events? In response, some digital experts have recommended that we intentionally reclaim the practice of deleting or erasing past data. Privacy expert Meg Leta Jones in her new book *Ctrl+Z: The Right to Be Forgotten* has argued that we need stronger legal options that require material to be deleted as well as proactive limitations on data collection.[31] As mentioned earlier, some European countries already have such laws and regulations and have also implemented legal regulations on the collection of data. These types of changes would limit and shorten the internet's long memory.

Similarly, Viktor Mayer-Schönberger in *Delete* cautions readers about the consequences of shifting from a human pattern of forgetting to a digital age of never forgetting. He recommends that we reclaim the virtue of forgetting. He writes, "Forgetting plays a central role in human decision-making. It lets us act in time, cognizant of, but not shackled by, past events. Through perfect memory we may lose a fundamental human capacity—to live and act firmly in the present."[32] He suggests that without the ability to forget, individuals and communities are forever tied to and even determined by past events, some of which are intentionally crafted and others a person may not have even chosen to share or reshare. Without the ability to forget, Mayer-Schönberger argues, individuals and communities are forever tied to and even determined by past events.

Past events shape us. They may even determine some aspects of our lives, either limiting or expanding options. Part of what the Christian tradition offers us is a way of approaching our past that encourages remembrance in a manner that is not determinative nor seeking erasure but is generative and takes account. We need to include in the vision of a "perfect memory" the many ways people change through acts of forgiveness. From a Christian perspective, change or transformation does not rely on forgetting. In fact, the opposite is true. Forgiveness and metanoia are based in accountability to our behaviors and past events, not a deletion of them.

Whether someone is self-authoring photos and comments, a friend posts something, or a media outlet publicizes a story, the repercussions can be devastating on multiple levels, including individual, family, and career. Thus, the need to cultivate digital awareness, diversity, and attunement as an ethical way of being is addressed in other chapters. In the meantime, while we work toward greater attunement and more reflective actions, as people of faith we need to consider our response to the archiving of digital data as well as the repercussions of actions and thoughts that remain forever via digital technology. How, as sinful people prone to personal, social, and institutional failures, do we react to a cultural shift that forever keeps track of those sins? In cases of moral transgressions, moral formation would be enhanced by focusing on forgiveness and metanoia rather than on whether or what information should be forgotten.

## MORAL FUNCTIONS OF FORGIVENESS AND METANOIA

These questions about the endless memory of the internet are especially complicated for Christians. While the digital cataloging of our lives often reinforces a notion of accumulation that can make it difficult for people to invest in change, the Christian call is a call of transformation—metanoia. We are an imperfect people striving to live Godly lives. Our inevitable failures require reflection and recalibration on a daily basis. This is both the gift and struggle of Christian life.

Christians approach forgetting differently. Is deleting information about past mistakes the same as engaging in the practice of confession or repentance? As Christians seeking forgiveness, are our sins ever forgotten or erased? When early Christians talked about transformation and the ability to radically change one's life and path, they often used the Greek word *metanoia*, which translates to "change of heart/mind" used to describe repentance or conversion. As discussed earlier, the notion of metanoia or repentance requires a truth-filled inventory of how one has behaved and believed in the past and sincere action toward change. Metanoia is not about forgetting. Such change is not only a feeling or thought; it was lived out in changed actions or reformed behavior. In that sense, metanoia is an ongoing process that must be continuously reaffirmed with an integrity of belief and practice.

Forgetting in relationship to metanoia would be an erasure when what is needed is an accounting. Mayer-Schönberger is, however, getting at a key aspect of the issue we face related to digital memory. Systemic change is needed in addition to personal change. Even when we correct our fiscal mistakes, take steps to repair relationships, and commit to ending harmful behavior, this personal metanoia falls short of reflecting on and impacting broader social change. Since the use of social media often perpetuates social inequality, metanoia must reach into social systems and ideologies as well as personal lives.

Exploitative social systems, such as sexism, racism, classism, and ageism, are held in place by ideologies as well as social practices. Returning to two examples from earlier in the chapter, we can see how digital technology affects the process of forgiveness and possible metanoia on a personal and social scale. For example, after TMZ recovered video surveillance from the hotel of the assault against his fiancée (now wife) and aired it online, Rice apologized publicly and pledged to change his behavior. With his fiancée by his side, he begged media outlets to stop showing the footage not because he was not guilty but out of respect for their privacy. Other stations picked up the released video and aired it as well. And as mentioned earlier, he was suspended and later removed completely from the NFL. In response, Rice has demonstrated sincere activity toward personal change, including counseling and outspoken advocacy against domestic violence. From what can be seen publicly of his actions and testimony, Rice has experienced metanoia.

Yet the traditional view of metanoia—moving away from one's own sinful nature and toward God—is shaped by a highly individualized concept of sin. It offers very little attention to corporate sins such as racism or sexism, in which we all participate unintentionally and often uncritically. Rice has not returned to the NFL despite his public apology and changed behavior. Some will say his return was based on a lack of skill; but we cannot be so naïve when numerous players since Rice have been convicted of similar crimes and suspended for only a single game. The viral visual of violence consciously or unconsciously impacted (and continues to impact) how Rice is and was treated by the public and thus by his employer.

Weiner, in comparison, has never publicly shown a commitment over time to change. In fact, the opposite is true. After the first incident, which caused him to step down from Congress, he was later found to have continued his actions as media outlets revealed in his run for New York governor. Even after that, he worked in political journalism for the *New York Daily News*. The point of this comparison is not to equate an act of direct, embodied physical abuse with the reception of unwanted genital photos. There is no need for an equation that calculates harm for either of these behaviors. Rather, Weiner as a White, well-educated, and well-connected politician did not suffer nearly the same public outrage or consequences. Granted, his marriage was not able to survive the third incident, which became known as a by-product of investigations of Hillary Clinton's email server during the 2016 presidential election. After further investigation, Weiner was also found to have texted sexually explicit material to a minor. It was only after that incident that he was held legally accountable. He was criminally penalized for his behavior with a twenty-one-month sentence in federal prison, the lowest possible sentence he could have received for his crime.[33]

Based on the colliding impact of social systems and personal choice, Ruether's understanding of sin and metanoia is helpful for those of us who are trying to figure out what Christian forgiveness looks like in a world of digital archives. Rice (and his wife) faced significant issues of personal and relational change. And Rice does so in a cultural context seeking a scapegoat for the NFL's and society's inability to stop sexualized violence. As many sports commentators have noted, the veiled racialized comments and depictions of Rice and his relationship evidence deeper systemic oppression related to who gets punished for violence against women, not to mention economic and sexist motivations behind the NFL's inability to uphold a zero tolerance policy on domestic and sexual violence. The notion of a zero tolerance exists in a similar structure of remembrance and forgetting that is bound up with digital data. Where is the call to change? Loss of job is the only motivator of behavioral change in a zero tolerance policy, and that has been shown to be ineffective. It is also counter to notions of relational accountability.

As a theological response, metanoia does not have the cross as its goal. There is no valorization of suffering for the sake of suffering. Ruether writes,

> One needs to move through anger to a deep enough self-esteem to forgive oneself and one's victimizers. To forgive, however, is not to forget, or to capitulate once again into victimization. It is only from a context of a certain confident autonomy, one that also allows some critical distance on one's own capacities to oppress others, that one can rebuild relations with others and with oneself, moving into increased capacity for mutuality.[34]

The metanoia of the oppressed is different than that of the oppressor, though both require rejection of the current historical and ideological structure of sin. We recognize the complexity that very few of us are either oppressor or oppressed; we are often both in multiple ways. In the role of the oppressor, one may need to move through shame and guilt rather than anger to recognize the need for forgiveness. Philosopher and law professor Jeffrie G. Murphy implores those who have been victims not to hasten past feelings such as vindictiveness or anger. In *Getting Even: Forgiveness and Its Limits*, he suggests that such emotions show us that something is amiss, that there is a need for soul-level change.[35] Or as Ruether would put it, these emotions are evidence of how alienation is affecting not only our relationship with the other but with self and ultimately with God.

For all those affected, accountability to the past—truthful recognition of harmful action—exists as a reminder of what not to do. Forgiveness is about a "change of heart," a consciousness raising at the level of the soul either about oneself or another, which takes into view social forces. Forgiveness often leads to penitence, partnered with a show of responsibility for past acts and commitment to change, but forgiveness does not require it. Forgiveness recognizes the potential (and actuality) of failure.[36] Forgiveness for Murphy is not to be hastened by passing over feelings such as vindictiveness or anger. He suggests that such emotions (what Ruether considers moments of social and personal consciousness raising) show us that something is amiss, that alienation is affecting not only our relationship with the other, but with self and ultimately with God. Metanoia happens in personal and social relationships in the midst of defeat, insufficiencies, and tragedies, as Ruether notes. Forgiveness is about (re)affirming self-respect, a turning toward the *imago dei* in ourselves and others as a way of establishing solidarity with others. She explains that once we see how structures of oppression impact our relationships with self and other, we need to reconstruct interpersonal relationships in mutual ways that disrupt systemic power imbalances through everyday actions.

## DIGITAL DISRUPTION

Forgiveness and consciousness raising are partners in metanoia that link personal, social, and ideological change. It is for that reason that we need to be more aware of the values embedded in the systems of dataveillance and social surveillance as part of an ethic of digital literacy. The archiving of data used for desire shaping perpetuates deformed economic and instrumentally driven views of each other and self. We need to unmask these distortions of relationship. This is an everyday process and perhaps every minute response on social media. As we push against alienation from self and others, so "we might begin to grasp anew what alienation from God might mean."[37] We can resist airbrushing our posts based on harmful cultural expectations, reconsider how we maintain friendship rather than accepting default settings, or disrupt the use of personal data for economic profit. These actions evidence the practice of attunement, dealt with in chapter 2.

Social surveillance practices and dataveillance can opportunistically reinforce systemic oppressions through personal actions. Metanoia happens when we actively disrupt that process and unmask these distortions of relationship. In our digital culture, one of the misleading claims is that dataveillance is harmless or at least morally neutral. Yet allowing companies to collect bulk information about online behaviors serves particular social structures of power—structures that perpetuate deformed economic and instrumentally driven views of each other and self. As users of social networks, we begin to prize measurement—likes, shares, and followers—over relationship. We too engage in and perpetuate the practices of surveillance. Peer-to-peer surveillance, called social surveillance, happens when we promote a comment or post by "liking it," post only positive information, review a friend's profile for past relationships or behaviors, or gang up on a blogger for comments counter to our political views rather than seek dialogue. This behavior not only means that social media users are both watching others and aware they are being watched, but it often maintains power differentials that reinforce offline inequalities.

Related to dataveillance we might disrupt the use of personal data for economic profit by deleting our personal search histories or permanently changing the settings to private browsers that cannot trace most online history. Of course, this is a personal response. We also need collective movements to change the way major institutional systems collect and use data. This will require legislative change. Organizations such as Privacy International seek to educate users, partner with civil and governmental organizations, and advocate for international laws related to data protection. Users need better control of their data, and this cannot be left to individuals to ensure because many of us do not even know the processes by which our data is harvested, used, or sold. Economic pressures usually keep government organizations from creating regulations that put users first, and this is exacerbated in countries around the world that have unstable governments, not to mention self-interest of governments that want to surveil their citizens. Education of users as citizens who can advocate for changes is a key aspect of Privacy International.[38] Not all internet businesses are using data for economic profit. There are nonprofit search engines, for example.

To counter the effects of social surveillance, we can resist airbrushing our posts in ways that further harmful cultural expectations around race, gender, and economic success narratives; reconsider how we maintain friendship rather than accepting default settings; and resist joining in comment or reply threads to reinforce our own opinion or denigrate another's. We can also amplify social media voices of people working for justice. Very simply, we can also publicly admit when we are wrong or apologize and take responsibility for our actions when we have made insensitive comments. Through these countercultural practices, we can begin to reshape social networks and digital media systems to recognize and affirm moral formation.

When we hold a richer view of sin, we reveal the role of socio-historical and ideological evils that are perpetuated by our use and the structures of digital media. One example of how the principles of metanoia might operate in the digital world is the site Fat, Ugly or Slutty, a user-run gaming website that holds online gamers accountable for their sexist and racist comments and actions. They publicly post user-generated messages that range from fat shaming to death threats against women gamers. It is a form of social surveillance that relies on dataveillance practices of collation and archiving. In an exchange chronicled by Lisa Nakamura in *Feminist Surveillance Studies*, a male gamer found his name published on the site. He asked to have his information removed because he says he was young, did something stupid, and would not like to "be reped [*sic*]—internet slang for represented—"that way."[39] He also notes that "I learned my lesson." The site moderator decided to remove his user name from the post but leave the information published. The response promotes accountability by not erasing the action, which supports metanoia on the level of personal transformation as well as continued social awareness that such behavior is morally wrong. This type of site uses data and social surveillance technologies but does so to combat social evils and hold individuals accountable.

As sinful people, our personal, social, and institutional failures go viral and remain archived for future generations to uncover. "To be forgotten" is a luxury in our digital culture, one to which Christians need not strive. Rather, we (and God) are better served by transforming our lives in an ongoing fashion evidenced by practices worthy of the forgiveness to which they respond. Metanoia rejects erasure of sin in favor of accountability on personal and systemic levels. The delete key does not and cannot forgive, but we can and do in ways that renew and restore relationship with others and God.

Metanoia extends to nonhuman aspects of creation. Our digital actions are not only virtual. Devices and networked systems provide a material infrastructure that comprise the landscape of the technological functions discussed in this chapter and others, including algorithms, datafication, dataveillance, and social surveillance. As we promote human difference, we must also consider biodiversity as part of the differences engineered into creation. We not only need to practice metanoia and attunement for each other and God; in chapter 4, we turn to cultivating a deeper sense of how our digital lives impact all aspects of creation.

## EXCURSUS 3: GOD AS PANOPTICON OR PRISONER

> For I was hungry and you gave me food, I was thirsty and you gave
> me something to drink, I was a stranger and you welcomed me, I was
> naked and you gave me clothing, I was sick and you took care of me,
> I was in prison and you visited me.
>
> —Matthew 25:35–36, NRSV

Some media theorists have likened the level of digital surveillance to
the panopticon prison building model designed in the 1800s, where a
single watchperson in a tower can see the whole yard of inmates. The
purpose of the panopticon was to instill a fear or anxiety of always being
watched. The same thing is done now by using digital cameras in prisons and all over society in stores, offices, and even our homes as well as
through the one-way mirror of our digital devices. It also reminds me of
the theological vision of God that I learned as a child.

The panopticon God who surveils humanity can lead to a disconnected and fearful relationship with God. God was always watching me,
keeping track of what I did, right and wrong (a little like Santa Claus). It
is also the opposite viewpoint to what we hear in the gospel of Matthew.
I read "I was in prison and you visited me," as a reference to the
incarnational presence of God as the prisoner, not God as the guard in
the tower.

Each semester, I have the opportunity to teach and visit with inmates, "inside students" as they are called in the Drew Theological
School's PREP (Partnership for Religious Education in Prisons) program. I am struck by how little the inside students know about digital
technology when they are perhaps one of the most surveilled populations in the United States. The government and in some cases private
companies that run prisons keep track of various biometrics, from fin-
gerprints to DNA, logs of movement through the system and during
incarceration, outside contacts, and so on. Additionally, many incarcerated individuals face significant disadvantages because they have no
access to learn these technologies. Withholding access to digital literacies is a key aspect of control over incarcerated people in the United
States.

In relation to digital technology, the panopticon is not as singular as
the watchtower metaphor might suggest. I understand it as the social
structures of society designed to benefit those in power, including the

corporations and ideologies profiting from the digital ecosystem. Most of us are unknowingly in various positions of being surveilled. There are gradations to the distance we are from the power center.

What changes in our relationship with God when we locate God as the inmate rather than a guard? Who or what, then, resides in the watchtower of the panopticon? Will we be judged on whether we show up for the prisoner? What transformations are required of social and political systems that use digital surveillance? What is our personal responsibility?

# 4

# CREATION CONNECTIVITY

**D**igital concerns are not only virtual concerns. As the number of devices and need for high-speed internet grows, so do server farms and
outdated electronics. Reduce, reuse, recycle is often associated with
plastic drinking bottles, not mobile phones, tablets, or computers. The
use and consumption of digital technology incurs significant costs to
energy sources and environmental resources. The rapid change in digital technology makes it almost impossible to follow the life cycle of
current digital technology, from manufacturing to disposal, not to mention all the energy necessary to power a device and applications in
between. Environmental concerns are digital concerns.

The current ecological crisis is impacted positively and negatively by
digital technology. Various technologies move us toward ecological solutions through more efficient and effective means of production, monitoring and reducing energy use, and moving us away from fossil fuels.
For example, smart home systems turn off lights when not in use, put
computing devices in sleep mode, and control the temperature of the
refrigerator or house to optimize cooling and heating systems. Colleagues meet in virtual conference rooms rather than having multiple
workers commute to one location. Digital technologies certainly offer
ecological solutions in a time of environmental crisis. On the other
hand, the minerals and metals that comprise digital devices are overwhelmingly mined using unsustainable practices that impact the earth,
the workers, and the economies of developing countries across the
globe. Though digital devices are efficient with energy use, they often

mask the collective amount of energy used globally to power these devices, which rely heavily on fossil fuels, not to mention the processing power at server farms contributes to significant heat production. It is a complex task to assess which and how digital technologies are reducing environmental impact or exacerbating it. We need to consider individual, communal, and systemic issues in concert to understand the full scope of the issue.

Many socially progressive Christian faith communities have reoriented their theologies and practices to creation care. Theological accounts have moved from interpreting dominion over nonhuman creatures, plants, and the earth to stewardship of the earth's resources. Faith communities move to become *green churches* with community gardens, clean energy use, and recycling programs. They relate local use of water to water shortages due to pollution and climate change affecting developing and underdeveloped countries. Some churches educate about the relationship between excessive food consumption and waste to the continuation of unsustainable farming practices and labor conditions. These educational initiatives attend to both individual and systemic awareness of the environmental crisis and humanity's role in it. Practices like these are undergirded by belief in the sacredness of the earth and all creation and emphasize interdependence.

Digital technology connects us with the earth's natural environment in ways we do not recognize and has a significant ecological impact. Yet many of us do not account for the role of digital technology in Christian responses to the ecological crisis. In the United States, digital technologies have an almost invisible environmental impact. In comparison to US use of automobiles that we see clogging highways on a daily basis, digital technology runs on what we think of as a virtual network so we do not see landfills full of old electronics placed in poorer, more industrialized areas or landscape and habitat disruption from burial of fiber optic cables or construction of signal towers. We need to cultivate greater awareness of how digital technologies affect the earth and our connection to this process.

Eco-theologians and ethicists provide insights for progressive Christians to understand the web of connectivity between social, ecological, and technological problems, enabling local action in response to what seems an intractable moral crisis. Digital technology functions, often in an unseen, cloud-like manner on deceptively tiny devices that belie their true ecological impact.[1] Similar to issues like greenhouse gases, we need to assess the full material impact of digital technology, including the process of mining for minerals, the infrastructure of fiber optic cable, and the geographic locations of server farms. The cloud computing of digital technology alters the earth's land structure far more than its atmosphere, though the atmosphere is also altered by the vibrations and heat produced by digital machines.

The vibrations that ripple from soil to air are a telltale sign of the ecological impact of digital technology that alters the language of the earth. To better interpret and hear this language, we must engage the digital literacies of "mattering"—recognizing the material impact of digital technologies—and "sensing"—increased awareness of the interactions between environmental forces. Aesthetic engagement, often experienced in and educated about through e-waste projects, enhances practices of mattering and sensing—further developing attunement and tying us more closely to creation. Shifting to theological models of interdependence with creation requires ethical responsibility as we develop ecologically friendly digital technology as well as alter consumer practices to promote a more just economy.

## LINKING ECOLOGICAL, TECHNOLOGICAL, AND SOCIAL ISSUES

The ecological crisis is as much a social crisis as a technological one. We are in a time of the Anthropocene, a commonly used term to describe the human influence on the earth, which has radically shifted environmental conditions. If humans created the current ecological crisis, then humans must also be partners in fixing it. Thus, this is the social aspect of the ecological crisis. Technological advances in travel, manufacturing, agriculture, and so on are the evidence of human meddling (or outright destruction) of the environment. As feminist theologian Rosemary Radford Ruether writes, "The capacity to be the agents of destruction of the earth also means that we must learn how to be its cocreators *before* such destruction becomes terminal."[2] Of course, not all people share the same benefits, burdens, or blame at the intersection of social, ecological, and technological forces. It is well researched that citizens in wealthier countries and wealthier citizens in poorer

countries use the majority of the earth's resources related to food, water, transportation, land use, and energy to name a few.[3] These are also
the same populations that are the most digitally connected. Human
action is the significant cause of the ecological crisis; exploring solutions
requires a nuanced evaluation of proportional effects and responsibility
by various individuals and communities.

So what of the use of digital technology in this process? Do we
abandon technological advance because of its contribution to the environmental crisis? In many cases, we do not know the environmental
impact of new technologies, especially digital technologies, until after
they have already begun helping or hurting the environment. Similar to
the approach of human intervention, technologies both help and hurt
the environment. Thus, a proportional evaluation of them similar to
human fault and accountability is necessary. On the other hand, some
researchers put total faith in technology to solve the ecological crisis
and argue for radical technological solutions that remove humanity
from the equation given how much damage we have already committed. To its detriment, this approach misses the influential connection
and interdependence that humans and technologies have with the fuller
ecology of the earth, which is also a living force, not simply an object or
objects to be manipulated.

The earth does not speak a language that fits modern oral and written forms of communication. There is a mathematical or scientific
rhythm to the earth's function that resonates with digital technological
innovation. And yet there is also a mystical responsiveness to the environment that is often accounted for by theological and spiritual writings. The earth cannot text us a map of energy use from a coal-burning
power plant across the country to the plug in our house charging our
mobile phone. The environment can only feel the impact of energy use
and respond within its own vocabulary of melting ice caps and rising
water levels. Without a direct impact, it is easy to ignore the earth's
speech acts, including the elements, plants, and animals that comprise
it. Many environmentalists suggest the earth does have a language; we
simply need to be retrained to interpret it. Learning the grammar of the
earth requires moments of conversation where humans slow down and
do less talking, writing, and texting, adding a complex level of communication to the scriptural insights about the Tower of Babel presented in
chapter 1.

*Gaia* is a term commonly used by eco-feminist theologians to name the subjective and active, even sacred, earth. For some, Gaia has no connection to a Christian divinity and may even be thought of as an antithetical power or presence. In response, Ruether confronts theologies that construct a rift between a patriarchal and other-worldly, masculine God of the heavens and Gaia. She suggests Gaia and God are in a conversation of divine creativity when she writes, "Her [Gaia's] voice does not translate into laws or intellectual knowledge, but beckons us into communion."[4] Other Christian feminist eco-theologians eliminate the divide between God's transcendence and God's otherworldliness and immanence. Eco-feminist theologians such as Sallie McFague describe the Universe of God's Body as uniting immanence and transcendence.[5] That is to say, God is in all aspects of God's creation. Eco-feminist theologian Ivone Gebara concurs, "Since we are a single Sacred Body, we are within the divine, and in a certain sense we are this divinity. The individual is not annihilated, but is instead related to a wider whole without which life would be impossible."[6] All of creation—flowers, animals, dirt, Grandma, trees, water—are part of the Sacred Body and thus divine. As Gebara notes, the sacred body can exist collectively, and individuals, plants, animals, and the earth's crust still retain their uniqueness. All contain their own being along with participation in the Sacred Body.

Our material interdependence with creation for air, food, water, and shelter is fairly evident, even when we abuse the relationship. Theological interconnectivity is a bit more abstract, yet no less real, and calls for an aesthetic relationship to creation that can be disrupted by and also enhanced by digital technologies. How do we begin to interpret the language of creation? Ecowomanist Melanie L. Harris begins her recent publication, *Ecowomanism: African American Women and Earth-Honoring Faiths*, by recounting her upbringing when her mother would sing and pray with the plants in their garden and around their house to nourish them. The vibrational awareness of creation's generativity is what Harris describes as "agricultural epistemology that is spiritual in orientation."[7] She connects this with the inseparability of spirit, nature, and humanity found in African cosmologies that sustained African American families like her own through the middle passage of slavery and in today's White supremacist US culture. In later sections, we will explore how earth's vibrations are connected with current digital technology use.

Following this ethical trajectory, digital users need to better understand the web of connectivity or links between social, ecological, and technological problems. Harris spends much of her text reminding the reader of the interfaces that connect White supremacy, capitalism, and political systems that have profited from the rape of the earth and Black women's bodies during slavery.[8] Her text demonstrates how anyone and anything can become expendable when we seek to dehumanize and desacralize parts of the Sacred Body. Similarly, Pope Francis in his encyclical Laudato Si' writes, "The human environment and the natural environment deteriorate together; we cannot adequately combat environmental degradation unless we attend to causes related to human and social degradation."[9] He does not recount historical crimes as boldly as Harris. Pope Francis does, however, call attention to how "billions of people become collateral damage"[10] to political and economic forces seeking profit. He also joins with many eco-theologians by noting, "we have to realize that a true ecological approach always becomes a social approach; it must integrate questions of justice in debates on the environment, so as to hear both the cry of the earth and the cry of the poor."[11]

Part of the intractability of the ecological crisis is its sheer size, leaving many Christians feeling morally incapacitated by the impact of their actions. Anthropocene is an acknowledgment of human power. "Recognizing anthropocene power acknowledges that managing earth systems as influential participants means learning to manage ourselves,"[12] writes Christian ethicist Willis Jenkins in *The Future of Ethics: Sustainability, Social Justice, and Religious Creativity*. He notes that it is not odd that a species could affect its habitat like humans have, but it is peculiar that a species "would do so knowingly."[13] It must be said that humans have knowingly impacted social structures as well. Domination and destruction is not only a way many humans have oriented themselves toward the environment but also toward other humans.

The shift from exploitation to sustainability using the anthropocene power is a central question that occupies many environmental policy makers, theologians, politicians, and educators. Jenkins notes that we regularly get "caught between ideologies of limit and of growth."[14] For example, the mining practices across sub-Saharan Africa used to harvest minerals to build mobile phones, which will be addressed in the next section, might lead one to argue for a massive limit on production because of the damage to the environment and workers' safety. Yet access to low-cost, reliable mobile phones is one of the most significant shifts in many African women's entrepreneurial path to economic independence, which contributes to the growth of gender equality and local stability. Limiting one affects the growth of the other. Obviously, there is not a one-to-one correspondence. Rather, there is a very intricate web of interconnection that affects far more social, political, and environmental issues than named in this example. And yet policy and practice often respond with a lack of nuance caught in a limit versus growth paradigm.

When faced with overwhelming, often intractable, problems that weave social, political, and environmental factors together, technology can be viewed as an objective solution. Trained by global capitalism and Western scientific progress, there is a tendency to view technology as an inherent good that spurs necessary progress. In *Global E-Litism: Digital Technology, Social Inequality, and Transnationality*, Gili S. Drori says, "ontologically, these digital technologies also carry the hope of bringing development."[15] The very existence or being of digital technology signifies development or progress. This is often why "technology, which, linked to business interests, is presented as the only way of solving these problems" when "in fact [it] proves incapable of seeing the mysterious network of relations between things and so sometimes solves one problem only to create others."[16] Disrupting a singular view of the function and end result of digital technology also requires admitting designers and users often cannot foresee the outcome of various digital technological design, as detailed in chapter 1, related to algorithms. No technology is neutral; it will and does "condition lifestyles" and shape "social possibilities."[17]

Digital technology development and use is a tangled ecological mess. Digital technologies are equally a form of e-waste as well as the primary mechanism for efficiently regulating energy use. For many of us, digital technologies and devices appear "green" because of their immateriality and efficiency. We are conditioned to "see" waste present in pollutants as concrete material substances such as plastic and Styrofoam. We cannot see the combined energy use that digital devices require or the earth-altering results of mining for minerals to manufacture digital devices. In the United States, to get people to care about carbon dioxide emissions, an invisible gas, the EPA gave it a quantifiable material structure by measuring and regulating it. US automobile owners, for example, had to have their carbon dioxide emissions tested when registering a vehicle. Sociologist Jennifer Gabrys calls this process "mattering"; it is a way to make an intangible both materially visible and have relevance or value.[18]

Awareness of the interconnectivity of the theological, social, ecological, and technological invites awareness of our location within the ethical issues at hand and opportunities for action in response. First, locating the divine in the Sacred Body resists dehumanization and desacralization as we turn to investigate the life cycle of digital technologies. Second, a greater awareness of the aesthetic is needed to understand the earth's language. And similarly a process of "mattering" can bring intangible digital waste into view for ethical evaluation. Such an evaluation requires fluency or increased literacy with digital technology manufacturing, use, and infrastructure.

## DATA MINING AND DIGITAL MATTERING

In a similar eco-feminist shift from a disconnected God in the heavens to an interdependent divine creation, we move out of the digital, immaterial cloud to the mineral connection of earth and digital media. In particular, the practices of mining connect digital devices to minerals and metals across the globe in addition to the social inequalities and violent unrest such practices have on societies. Second, we follow the energy trails of our digital devices from the minerals that make them function through the wireless spaces that connect them. The material life of digital technology is heavy and wasteful in a way we have yet to morally account for in ecologically just ways.

Anthropocene names the human power to alter the environment but only hints at the moral consciousness of human intentions. When Jenkins notes that humans are the only species to "knowingly" alter their environment, his statement suggests humans have and are aware of their ethical responsibility. Yet the sheer size of the crisis often results in a stuckness—the individual desire for change up against the magnitude of a global crisis. And yet individual action taken collectively can have a significant impact. That is in fact the contours of action that led to the crisis in the first place. Global consumption, disproportionately led by Western developed countries, borders on the obscene. Media theorist Jussi Parikka has coined the term *Anthrobscene* to not only name the role humans have played in altering the environment but also to clearly signal and morally condemn the involvement as "obscene."[19] Parikka argues that those in power have knowingly perpetrated ecological disaster on all of humanity. The price of connection is a cost to the earth that in turn socially, politically, and economically affects those of lower socioeconomic status, often in rural areas in the United States and globally, and disproportionately the poor in mineral-rich countries. Those of us not affected firsthand by these environmental and economic issues are often unaware. So we trace the connections and their costs.

## Digital Is Earth

Computing devices in general and mobile devices in particular are manufactured using a wide variety of metals and minerals. The making of a mobile device can require thirteen mineral commodities from seventeen countries, ranging from the United States and China to Peru, Rwanda, and the Congo.[20] Those minerals evidence the "material histories of labor and the planet."[21] Much of the mining takes place in developing or underdeveloped countries with little enforceable environmental regulation, labor law, or political stability. "Data mining might be a leading hype term for our digital age of the moment but it is enabled only by the sort of mining that we associate with the ground and its ungrounding. Digital culture starts in the depths and deep times of the planet,"[22] writes Parikka. Digital devices are literally composed of earth from around the world.

Mining practices for minerals has significantly affected a number of countries. According to Congolese theologian Muhirgirwa Rusembuka Ferdinand, "This situation [in the Democratic Republic of the Congo] can be largely attributed to the fact that companies take advantage of the weakness of the state's regulatory powers. The companies gain a financial advantage from avoiding the costs of environmental cleanup, and the legal and political enforcement mechanisms allow those companies to externalize the costs of mining. Downstream ecosystems and communities bear the costs instead."[23] Environmental practices that degrade and erode the earth mirror how workers and local communities are often treated. The harvesting of conflict minerals has led to genocidal war in the Congo. "Thousands, including many children, have died in the coltan wars, and ultimately those lives are lives lost to our electronic pleasure."[24]

When it comes to mining lands and resources, laborers—the indigenous who inhabited the land—and even the local governments often do not profit. Leaders of foreign companies and individual officials profit as do the majority of digital device consumers who want newer, faster, and less expensive devices. South African theologian Peter Knox, SJ, argues that "patterns of ownership often follow patterns of colonial conquest and privilege."[25] The owners and exploiters of land are less likely to be Western countries that colonialized the global South; now, they are Western companies dominating global markets. In addition to labor practices, sustainable mining may not be possible. Knox notes that rare earth elements, many of the minerals necessary for digital devices, are neither renewable nor in endless supply. Then, there is the amount of water and electricity needed to run the mines and extract the minerals.

To respond to exploitation of people and the earth, Knox suggests that we reconceive of how we grant ownership of land and the responsibilities that come along with it. Companies claim ownership of land to exploit it; consumers see the payment for their devices as a sort of payback for their ownership of these minerals in a new form. The costs do not add up. Common use of land approaches need to outweigh ownership models so that local people and sustainable practices are at the forefront of mining reform. And even if we are to permit ownership, then we need strong civic and governmental groups to enforce concomitant responsibility to use land in a manner that breaks cycles of poverty that exploit humans and natural resources. Likewise, Ferdinand suggests, "To promote good governance in the mining sector will require that all stakeholders work on specific causes locally (transparency, accountability, fight against corruption) and globally (fair trade agreements, implementation of corporate social and environmental responsibility)."[26] Ferdinand balances individual and systemic responses as well as local and global responses.

Mining is not the only ethical issue in the manufacturing process. "After extraction, the assembly process of electronic devices likewise often takes place under horrendous conditions."[27] Ongoing news stories appear related to worker issues in Apple's assembly plants in China, for example. Inhumane standards of work range from exposure to chemicals to long overtime hours to extreme surveillance of workers. When we link the manufacturing process to a device, we find that digital devices are material evidence of poor labor practices and planetary degradation.

Digital devices are designed to be quickly phased out as technology changes rapidly and manufacturers want to sell new items. Marketers appeal to desires for faster, flashier, and larger capacity in smaller containers. As consumers throw away, or hopefully resell and recycle what can be reclaimed from digital devices, electronic waste piles up. "The earth, our home, is beginning to look more and more like an immense pile of filth."[28] Most of these piles are hidden from those with the wealth to purchase ever-newer digital devices. Pope Francis describes the morally malformed consumer relationship when he writes, "These problems are closely linked to a throwaway culture which affects the excluded just as it quickly reduces things to rubbish."[29] In the United States, many poor urban and rural communities absorb the digital waste of the primarily White, educated upper class. Across the globe, many developing nations are erecting skyscrapers of industry as quickly as landfills grow within and around poor communities.

The earth is integrally part of our digital devices. From the first elements that comprise a digital device, it is connected to global geography. Parikka writes, "The geological material of metals and chemicals get deterritorialized from their strata and reterritorialized in machines that define our technical media culture."[30] As the metals become our mobile devices, they in turn "enable technological mobility."[31] Consumers carry small pieces of Africa and China around in their pockets. As users, we often remain unaware. Yet these technological assemblages are evidence of this transformation of energy.[32] It is time we start paying more attention to how these assemblages form and where the energy transformation leads. A just, moral response requires a collective will that functions on individual, community, and structural levels. We can begin developing this collective by disrupting the consumer orientation toward digital devices in a way that shakes us out of moral complacency, mattering these connections, making us materially aware of them, infusing them with social and moral value. The valuation increases the likelihood of legibility by humans.

## Mattering

Digital culture exists because of its reliance on forms of energy. This superhighway is not populated with carbon-emitting cars, but it does produce a significant amount of pollution in the form of energy use, transfer, and waste, from individual devices to fiber optic cables and data centers to server farms that rely on unclean energy sources that emit their own energy by-products. Gabrys is helping to show how this "matters." She writes, "Indeed, if one were to account for all the energy used to manufacture, power, connect, cool, maintain, and eventually recycle and dispose of electronics, estimates of electronics-related energy use would increase even further."[33] Most energy assessments of digital technology focus on the production and use of devices. We may consider the energy that runs production plants or stores that sell our electronics, but mostly, we focus on how much energy a small device uses in our local setting. That amount is minimal in comparison to other energy uses. Quantifying energy use for a variety of digital technologies awakens the user to a wider ecology of energy use.

The material costs of providing power for digital technology is environmentally significant. We may assume the digital infrastructure "operate(s) seemingly free of resources. In another sense, the energy required to power electronics results in distinct forms of pollution that are different from the stacks of abandoned digital devices often associated with electronic waste."[34] Power grids rely on fossil fuel–burning energy plants. Data centers and server farms require a tremendous amount of energy for processing and cooling. These servers are needed all over the earth to power the information economy. Their construction and vibration have a terraforming effect. In addition to servers, fiber optic cables need to be laid to grow the digital communication framework.

In looking only at the financial industry's use of algorithms for trading, MIT Media Lab and technology entrepreneur Kevin Slavin notes that microseconds make the difference, and thus, companies want to be as close to servers as possible to minimize the time it takes for trades to take place.[35] In a global financial system, that means servers will eventually need to be placed throughout the world, including on or in the oceans.

Parting waters for server placement sounds biblical in proportion. That's not exactly what happens when a data center is planted in the ocean. For example, Microsoft has deployed a data center on the Pacific Ocean floor.[36] They hope in the future to deploy more. The project is called Natick; the data center is self-contained and sealed. There are computational benefits in terms of speed for those close to these data centers. There are some ecological benefits as well. If done correctly, these types of data centers will use less energy for cooling because the ocean floor provides a cooler environment that distributes heat more efficiently. Tidal energy could also be harvested to provide power to the centers, thus reducing fossil fuel use by data centers placed on land.

Does the use of cleaner energy to run the data center balance out the possible environmental effects to the ocean floor? At the bottom of the ocean, the loud hum of the data center cannot be heard by humans.[37] The constant hum of machines often affects workers' hearing and can disturb the natural environment surrounding these primarily land-based facilities.[38] A visit to a data center materializes energy in a very concrete way; one can actually hear the digital information infrastructure speaking. Some music groups have recorded the hum for soundtracks.[39] How will the ocean floor or creatures respond to this new form of communication that emits a hum? What might be the shifts in nature's structure as energy is released and captured by a data center at the bottom of the ocean? We do not know the answer because we cannot foresee the consequences of what at first might appear to be positive ecological benefits of technological advance.

We have to attune ourselves to recognize the earth's language. This language is aesthetic in quality, which humans have for some time dismissed as a lesser form of communication, opting instead for the written and spoken word. Applied mathematician and scientist Samuel Arbesman in *Overcomplicated: Technology at the Limits of Comprehension* sums up the anthropocene and digital connectivity of creation in this way:

> Formal title or not, the relationship between our human-made systems and the natural world means that each of our actions has even more unexpected ramifications than ever before, rippling not just to every corner of our infrastructure but to every corner of the planet,

and sometimes even beyond. The totality of our technology and infrastructure is becoming the equivalent of an enormously complicated vascular system, both physical and digital that pulls in the Earth's raw materials and emits roads, skyscrapers, large populations, and chemical effluent. Our technological realm has accelerated the metabolism of the Earth and done so in an extraordinary complicated dance of materials, even changing the glow of the planet's surface.[40]

Once materialized in forms and structures legible to humans, energy use by digital devices, server farms, fiber optic cables, and data centers *matters*. The planetary hope is that we will moderate the use and growth of digital technology in a way that pays attention to the connectivity of humans, earth, and technology. As we hear the hum of Harris's mother singing to plants, so too the vibrations of a server farm speak to nature around it. The ecology of the digital technology infrastructure is as much part of the earth as the earth is part of it.

## RECONNECTING WITH COCREATIVE RESPONSIBILITY

Whether we are moving theologically from heavenly cloud to divine earth or ecologically from digital cloud to oceanic hum, material mattering of digital technology evidences the interconnectivity of creation. With this recognition comes clarity of ethical response. First, individual awareness of connection with the earth and all that is sacred must be balanced with selves increasingly composed of data. Second, the desacralizing of the creation through environmental degradation must be linked to the dehumanization of poor—often Black and Brown—bodies, globally. Much of the current digital infrastructure is built on the destruction of land and people across the globe. Third, and perhaps most important, such destruction reduces biodiversity by removing nonrenewable minerals and metals, damaging ecosystems on land and below the sea, and harming communities through poverty and war.[41] While this claim may seem extreme given many environmental and social benefits of digital technology, it is also true.

Digital infrastructure is built on "repetitive cycles of consumption, built-in obsolescence, poor resource use, and labor inequalities, in addition to environmental pollution."[42] So we must find ways to alter this cycle if we have any hope of actual digital progress—that is, progress that honors interconnectivity and interdependence to promote biodiversity rather than destroy it. Digital infrastructures must be an aspect of how we understand biodiversity. The full ecology of digital life is interdependent with the earth and all its life forms, especially humans. Biodiversity is generally understood as the variations of life—plants, animals, organic materials—in an ecosystem. The greater the diversity, the healthier an ecosystem is. Digital technologies are in a coconstitutive process that radically reshapes humans and the earth. Its active role not only changes the variety of life in an ecosystem; it can be understood as one of the life forms. The throbbing of data centers, the reterritorializing of minerals, and algorithmic seismic shifting are organic in their impact and growth.

Adding digital infrastructure to biodiversity is not a mechanical way to increase biodiversity in the midst of global land, plant, and animal loss. Rather, continuing to view digital technology as a machine or tool separate from the created order dismisses the reality of how deeply the digital infrastructure is woven with social and ecological life. Biodiversity is negatively affected by climate change, pollution, and habitat degradation. Just as humans harm their own ecosystem, so too does digital technological expansion harm its own sustainable future.

Currently, we live in a state of biological impoverishment as Willis Jenkins claims. He defines biological impoverishment as "degradation of the systems that make earth a habitat for humanity."[43] Centrally, Jenkins is concerned with how we can come to value biodiversity in order to move out of biological impoverishment. One of the central problems with valuing biodiversity is that humans routinely undervalue the "true costs of ecological resources and services,"[44] such as the amount of paper it takes to write, edit, and publish a book. We never pay back the level of resources we take. Going paperless is not the solution. The digital infrastructure that allows for book publishing from word processing, emailing, copyediting, design, and layout equally uses resources of energy and materials in a nonrenewable manner. The tree is as much a part of biodiversity as the energy, minerals, and metals of digital technology. Valuing biodiversity means recognizing and cultivating interdependence with creation rather than exploiting it.

Humanity's role as cocreators is a generative location for moral action. Valuing biodiversity is a process of honoring the sacred in creation.

"Creaturely diversity reflects in knowable ways the unknowable Creator . . . biological loss impoverishes the human ability to know God, which Christians hold as humanity's deepest desire. The point: within the right sort of practices, earth can teach humans how to value it and perhaps even offer a taste of infinite wonder."[45] This will require a significant reshaping of human desire, which digital technology is already affecting. The question is whether the constraining, shaping, and framing of human desire will continue to be in service of current systemic structures of power, such as global capitalism, which promotes the destruction of some for the benefit of others, or toward ecological economies based in biodiversity led with cocreator accountability to promote the flourishing of all. "Practices that slow the metabolism of consumption can begin to nurture a different economic ethos,"[46] writes Jenkins. While he is referring only to human practices in his writing, I would add digital technologies have their own metabolism of consumption that needs to be slowed as well.

## Sensing

One human and digital connection that is moving toward a recognition of and limiting of consumption for the benefit of the earth is the Citizen Sense Project. Gabrys, whose work on mattering has already been presented, began the Citizen Sense Project in 2013. Funded by the European Research Council, "the project investigates the relationship between technologies and practices of environmental sensing and citizen engagement." There are three project areas: pollution, urban, and wild sensing areas. Digital sensor devices are often used for scientific study of environmental issues. This project harnesses the way in which "practices of monitoring and sensing environments have migrated to a number of everyday participatory applications, where users of smart phones and networked devices are able to engage with similar modes of environmental observation and data collection" to promote wider engagement in environmental awareness and action.[47] Everyday citizens use smartphones and networked devices to collect data and share observations.

The sensing projects are not simply ways for local individuals to gain scientific skills in understanding environmental impacts. The purpose draws together ecologically helpful modes of digital technology use that shifts people's desires to better understand, care for, and eventually respond to environmental concerns. Projects include sensory devices such as moss cams, air-pollution monitors, soil sensors, and smart-building energy sensors. "Sensing" is a digitally enhanced form of singing to plants that deepens the interdependence of people, earth, and digital technology while building awareness on individual, community, and structural levels. Citizen Sense asks the question this way, "In what ways do the political, material and affective orientations of sensing devices fulfill their performative potential, and to what extent do the diverse practices of citizens' everyday sensing practices reorient the intended programs of these devices?"[48]

In Gabrys's extended research of the environmental use of sensors as computational technology, she argues that sensors create techno-geographies.[49] That is to say, sensors, for both their computational ability and their material constructions, become part of the ecology of the spaces in which they reside and shape the environment. Sensing shifts human interaction with environments from two-dimensional spaces we map to "a field of resonance and relation that can be drawn into and materialized in the experience of subjects."[50] Gabrys writes, "these programmed environments draw attention to the tunings and attunings that occur through computational and environmental operations. Ways of feeling and accounting for environments and environmental relations are activated or otherwise delimited through computational sensors."[51] For example, the information provided about the causes of air pollution in an area changes how we define the air, how we behave, and what policies we may advocate for on a community level. The sensor is a key aspect of change in the relationship between the people and their environment.

We ought to be cautious of romanticizing such technology as always helpful and consider in what ways a programmed earth becomes a surveilled earth. Gabrys addresses these issues as well. In the end, the Citizen Sense Project affirms what many environmentalists suggest, that an aesthetic awareness disrupts the human consumptive desire for more. Or as Jenkins might argue, we reform our desire for more to a "right mode of 'more'" by reshaping our perceptions to want "more justice, more beauty, more dignity."[52] Aesthetic connections that come from computational sensing may contribute to this desire-shaping effect.

In this process, we must specifically and directly address consumptive habits of humans, especially related to digital technology. More sensors means more devices, more energy, more mining. Gabrys is not arguing for this nor am I. Rather, the above example is one way to engage digital technologies and articulate the ways they contribute to biodiversity in ever-expanding ways while recognizing the networks and availability that already exist. Other aesthetic alternatives may come from directly engaging communities about e-waste. "Although more environmentally benign technologies certainly have a major role to play in achieving sustainability, technology alone will not be sufficient to save us. Also direly needed are changes in lifestyles and in moral values, including a humbler relationship with nature and an embrace of sufficiency rather than the consumerist ideal of ever-expanding consumption,"[53] writes Christian ethicist John Sniegocki.

A different sensory awareness needs to be cultivated as well as following the paths of our material use. As cocreators with God, we need a deeper recognition of the material effects of development. This can lead us to assess whether "more" is in service of beauty, justice, and dignity. Interconnection with the Sacred Body has to be cultivated because many humans have spent significant time and energy distancing themselves from this awareness. Cocreational responsibility requires knowledge from a variety of sources to fully return to such an awareness. Sensing is one approach.

## Art of E-Waste

The emergence of e-waste art is not necessarily a new phenomenon. For decades, artists have used recycled materials and garbage to create art that raises awareness for environmental concerns. The art of waste, as many have named this process, invites the viewer to "rethink your love affair with the exploitative and toxic tools of prosumerism."[54] E-waste art ranges from art that reclaims the refuse of digital use to display the mass amount of consumption, such as circuit boards, phone chargers, and obsolete mobile phone models, to photography and installation art that chronicles the life cycle of devices or shows the contents of animal stomachs filled with plastics.[55] It also includes the aforementioned techno-music movement, which integrates data center–generated or even household-generated digital noise. The aesthetic invitation of these artists is directly linked to education, awareness, and action.

As digital technology use expands across the globe, there are more and more digital devices that are designed to be replaced after two years. There are also all the accessories, accompanying printers, monitors, consoles, covers, and so on. Artists use a variety of means to make this waste intelligible. This is the waste that can be seen. A plethora of chemicals are used to build digital devices, from the mining of minerals to the disposal of components.

> Some of these toxins affect the workers who assemble devices, some can impact users, all impact those who disassemble devices (especially when components are burned), and all of us, though hardly equally, are impacted by landfills piling up obsolete e-devices that can leach into local water supplies and agricultural lands, as well as incinerators that send this material into the air where they can potentially drift over any community.[56]

In addition to intangible pollutants, the disposal of e-hardware ends up in less developed parts of the world or communities in the global north where many poor, often people of color, live, including Native American reservation lands. "Most of the toxic waste from electronic devices ends up leaching into the land, air, and water in the poorest parts of Africa, India, China, and other parts of the less industrialized world."[57]

E-waste artists are "reinterpreting the 3-Rs—reduce, reuse, and recycle—with thought-provoking, whimsical, and even utilitarian creations."[58] In the process, they are reeducating consumers. Environmentalist Annie Leonard in 2007 released a video called the *Story of Stuff*, which became a YouTube phenomenon and educates consumers about the full production of "stuff." She names the cycle "take, make, waste" to correct the lack of a practice that actually seeks to reduce, reuse, recycle.[59] Rarely do e-waste artists use their creative mode outside of a digital media landscape. Instead, they pair their creativity with digital outreach, such as the *Story of Stuff* video on YouTube. Similarly, Molleindustria developed an app called Phone Story in 2011 that takes players through the digital life cycle of their mobile phone. The game has four areas that teach the user about mineral mining in the Congo, labor practices in China, e-waste disposal in Pakistan, and consumerism in the West or global North. The developers say, "Phone Story is a game for smartphone devices that attempts to provoke a critical reflection on its own technological platform. Under the shiny surface of our electronic gadgets, behind its polished interface, hides the product of a troubling supply chain that stretches across the globe."[60] They use their profits to support educational and policy-based groups working for educational change as well as workers who are affected by these practices.

Chris Jordan is another example of an artist raising awareness of global consumption. He shows his art around the world as well as on his website, "Art Works for Change," where he organizes the viewing experience around five principles—delve, learn, interpret, reflect, and act.[61] In a section related to "delve," between a photograph of a recycling yard and the unaltered stomach contents of a Laysan albatross fledgling, Jordan's own words are featured:

> As an American consumer myself, I am in no position to finger wag; but I do know that when we reflect on a difficult question in the absence of an answer, our attention can turn inward, and in that space may exist the possibility of some evolution of thought or action. So my hope is that these photographs can serve as portals to a kind of cultural self-inquiry. It may not be the most comfortable terrain, but I have heard it said that in risking self-awareness, at least we know that we are awake.[62]

Amid his photographs, he shares additional resources for site visitors to learn more and reflect on their own habits. He then invites readers to a featured action of pledging more environmentally healthy practices. Each practice links out to more art that depicts the problems as well as information on how individual and community practices can change in addition to environmental policy action. Still other artists reclaim materials turning devices and parts into jewelry, furniture, clocks, or home décor, which creates everyday reminders of the permanency of the life cycle of digital technology.[63]

These projects move us beyond the overwhelming apathy generated by global production and consumption cycles to call forth cocreational awareness but also a spiritual engagement and connection. Desire reshaping may be in service of new, less earth-destroying habits. It is one of the ways, as Pope Francis suggests, we "continue to wonder about the purpose and meaning of everything. Otherwise we would simply legitimate the present situation and need new forms of escapism to help us endure the emptiness"[64] The art of waste is not a form of ethereal escape into digital clouds away from tangible earthly connection. It is an ethical practice that brings us closer to the Sacred Body, searching for a biodiversity that can sustain the planet, our spiritual lives, and digital connections.

The interconnectivity of humanity, digital technology, and creation is often felt more than discussed. Whether through digital literacies we learn to make digital ecology intelligible, we are still interdependently bound. The theological task of increasing biodiversity to strengthen the Sacred Body requires we seek out and honor the divine in all of creation. Aesthetic practices such as sensing and visual and tactile artistry are examples of "mattering" that seek to return the sacred to creation and humanize the people who become collateral damage within the global digital infrastructure. Harris states, "in addition to recognizing the links between social justice and earth justice, ecological reparations problematize some of the frames of environmentalism, acknowledge the impact of colonial ecology, and replace dualistic understandings that divide the earth from the heavens, for example, with a more fluid frame that values interconnectedness and interdependence."[65] Critical digital literacies call forth a new ethical aspect in these reshaped relations.

Awareness must lead to new ethical practices if ecological reparations—material compensation to the earth and peoples most directly affected by ecological degradation—are to become part of Christian justice-seeking movements. Digital technology as part of biodiversity requires as much nuance in the promise of technological fixes as skepticism. If we forefront the locations and peoples most affected by environmental digital destruction, we can move beyond the limit versus growth standoff. In that process, we "develop practices for learning to desire real wealth through efforts to overcome impoverishment."[66] Central to the work of "mattering" is an ethical shift in value—value as moral fortitude, spiritual connection, environmental impact, and biodiversity. Awareness about the digital environmental impact is an important aspect of digital literacies. The social, economic, and technological systems we live within, often benefit from, and reproduce in our everyday actions are evidence that we value independence over interdependence, destruction over creation, and alienation over relationship. This is not the case for everyone nor need it be for anyone. We, however, will need to disrupt the search for easy technological fixes and instead cultivate digital diversity within the fuller expression of biodiversity as a Christian moral responsibility.

## EXCURSUS 4: DIGITAL CLOUDS AND DIRT-FILLED DEVICES

Be still, and know that I am God!
I am exalted among the nations,
I am exalted in the earth.

—Psalms 46:10, NRSV

Being still in a digital age is difficult. Waiting in line or waiting in a traffic jam is now an opportunity to connect, to get work done, or to play a game on a smartphone. The whir and whiz of digital technology vying for our attention is not about focus; it is about distraction. The distraction of entertainment, connection, or efficiency keeps us from moments of wonder, boredom, flourishing, and stillness.

I find in Jesus' incarnation a reinforcement of human particularity *and* divine presence. Sallie McFague in *The Body of God* writes, "God is not present to us in just one place (Jesus of Nazareth, although also and especially, paradigmatically there), but in and through all bodies, the bodies of the sun and moon, trees and rivers, animals, and people. The scandal of the gospel is that the Word became flesh; the radicalization of incarnation sees Jesus . . . as a paradigm or culmination of the divine way of enfleshment."[67] God comes to humanity as a person, taking on the limitations and delights of a body. At the same time, God is also in all aspects of creation.

God is more expansive than I can comprehend. As the earth, minerals, and metals are present in my digital device, is God? As the bodies of my friends and loved ones are present on my social networks or video connections, is God? I find myself answering, "Yes." My relationship with God is always mediated by something or someone—scripture, another person, a lit candle, the sun on my face, and so on.

Being still, to know God, involves conscientious awareness of my relationship to digital technology. I don't mean that I use more devices or am always connected. I mean I have had to reorient myself to engage digital technologies from the perspective of flourishing. Flourishing calls forth in me a sensibility that includes my embodiedness; my emotions; the aesthetics of digital technology; the beauty, power, and fragility of the earth; and the humanity of the people to whom the technology connects. Flourishing is a mutually, interdependent endeavor.

When we take a picture, can we enjoy the beauty in front of us *and* take a second to be grateful for the ability to capture that beauty, to remember it days from now, or to share it with others? How will we remember the beauty that is destroyed in the process of manufacturing the device that took the picture? Reflection is not meant to seize or stifle everyday living. It is meant to propel us toward awareness and accountability.

How might you know God differently in and through digital technology? What does flourishing in a digital world look like for you and for the earth?

# 5

# ETHICAL HACKING AND HACKING ETHICS

Hacking has a morally negative connotation. However, hacking as a practice can be enormously helpful to making systems stronger, raising public voices in a landscape dominated by corporations and governments, and adapting software and hardware to new purposes. The technical aspects of hacking are often considered the skills of the digital elite, something that separates average users from the digital expert. Everyday users also engage in hacking in a broader sense, and that is the focus of this chapter. Bringing an ethical awareness to use of digital technologies necessitates we adapt, repurpose, or even disrupt the current digital ecosystem. This is what I mean by ethical hacking.

Hacking relies on a process-based approach that seeks social change through creative engagement. The outcomes of digital expansion and adoption are not predictable, and thus our ethical evaluation must also be nimble and fluid. Such engagement often results in participatory cultures built by online users who share social interests and ethical aims. This can be a community of activists or a group of fans. Either way, they are engaged in a creative act of adapting and modifying technologies to fit their community need. Hacking, in tech circles, is understood in relation to computational skills used to access a digital technology through vulnerabilities that then can be patched once found. In this chapter, I expand hacking to include the ethical call to gain access to the ecosystem of digital technologies and define the vulnerabilities to be patched as the perpetuation of social inequalities and injustices.

Throughout the previous chapters, I have focused on specific aspects of the digital ecosystem that, once described and understood, increase the reader's digital literacies. Digital literacies include facility with technologies as much as they do recognition of the changes and challenges that the particular technology poses to social practices, identities, and relationships. This chapter follows a familiar rhythm, with one deviation. I begin with the scriptural metaphor of cultural change offered in Isaiah that recounts the altering of swords into plowshares (Isaiah 2:4). I will not provide an exegesis on the scripture, so much as explore the ethical approach needed to engage in the kind of cultural change alluded to in the passage. And the ethical approach articulated in this discussion brings together the exploration of difference, attunement, metanoia, and cocreation from previous chapters, rather than proposing a new theological resource. When we practice these moral approaches in a digital world, we are engaging a form of hacking that relies on transformative and liberative digital literacies. We do this in participatory communities that provide a feedback loop for moral growth.

## SWORDS INTO PLOWSHARES

What is required to transition a military tool, such as a gun, into a civil, agrarian tool for food production? The question alludes to both the complex metal work needed and the cultural reorientation of humans to the technology. Repurposing, or hacking technology, also shifts uses of and social orientation to that technology. This is a creative act that is both responding to an ethical claim and hoping to forge new ways of ethical living. There is clearly a social justice aim to the process that leads users to assess the moral formation that technological tools have on us, our environment, and our resource investment in them. The Isaiah passage is about God's judgment of nations and the ways in which war has destroyed the strength and growth of Jerusalem: "He shall judge between the nations, and shall arbitrate for many peoples; they shall beat their swords into plowshares, and their spears into pruning hooks; nation shall not lift up sword against nation, neither shall they learn war any more."[1] Throughout the text, Isaiah is calling Jerusalem to "just social practices."[2]

RAWtools is an organization that has taken the scriptural call of Isaiah seriously. They invite people to send them their guns, and then they repurpose that gun into a gardening tool to return to the sender. The motivation is more than transitioning metal from one tool to another. In describing their work, they say, "It's not enough to just make a lot of tools from guns. We need to help teach each other new ways to solve our problems through relationship, dialogue, and alternative means of justice."[3] Whether Jerusalem actually heeded Isaiah's prophetic call, like RAWtools is doing, we do not know.

RAWtools asks communities to "dare to use our imaginations to change our impulse." They believe shifting the way we use technologies will have this effect. "Beating swords into plowshares, spears into pruning hooks, and guns into garden tools creates a dynamic shift in our investment in time and resources. . . . It doesn't mean we are all to become gardeners, but it does mean we can invest in providing life sustaining resources for our communities."[4] Shifting the technology of the gun from a machine that can kill or wound to one that produces life through gardening affects the imagination of the user and opens up a variety of resources for the community.

The prophet Isaiah is "well known for his insistence that Jerusalem's elite bear deep ethical responsibility for those they govern."[5] Isaiah expresses a proportionalism in his assessment of who has moral responsibility. We may affirm a similar assessment when it comes to the morally problematic ways digital technologies impact individuals, society, and the earth. However, similar to the argument of RAWtools, even when governments bear a significant responsibility for war, individuals in a society are also responsible for the broader cultural, legal, and political support that maintains the ubiquity of the technology of guns. Thus, we need strategies that lead to individual, communal, and systemic changes.

The internet's origin is also bound up with military purposes. A researcher from Massachusetts Institute of Technology (MIT), one of the first to have described a way of networking a group of computers in the early 1960s, was the head of the computer research program of the Defense Advanced Research Projects Agency (DARPA) for the US government.[6] The work of his team was to figure out a way for computers to network, thus more easily sharing research and communication. These details are not meant to insinuate that the internet is synonymous with a gun because both technologies are a primary domain of the military. Rather, it is to remind us that the internet is a product of a government project, directed by a Defense Department unit and in cooperation with government-contracted researchers at outside universities and businesses. The general drive to network computers was about accessing computers at other locations that could run processes faster, not necessarily for people to communicate with each other. Development of any technology has intended uses and unintended consequences. In this case, the ability to communicate, share research, and access computer processing power is generally speaking a good outcome. Unless of course, the purposes of communication, research, and processing power lead to control of information, citizens' activities, or consumer exploitation.

Once the technology was successful, the community of researchers and the number of networked computers grew. "The Internet is as much a collection of communities as a collection of technologies, and its success is largely attributable to both satisfying basic community needs as well as utilizing the community in an effective way to push the infrastructure forward."[7] The communities expanded with significant relationships to commercial industries. The internet as a techno-cultural tool of communication shifts human ways of doing, thinking, being, relating, and making meaning. The diffuse and yet networked nature of the internet's design means there is not one primary intended use for this tool, such as there is for a gun. The medium of the internet and the digital technologies that grew from it require a focus on the formation of communities that come together via its use, how they reshape and grow its basic technologies, and the functional affordances (benefits) and constraints (limitations) it allows in order to assess ethical questions.

In the early development of the internet and computers, very few Christian social ethicists wrote about computational technology. There is one exception that I could find. In 1986, Christian social ethicist Edward Le Roy Long Jr. penned the essay "The Moral Assessment of Computer Technology" in *The Public Vocation of Christian Ethics.* His essay focuses on the impact of computers on the workforce and the home. The essay prioritizes social observation and review of scientific insights to describe the shifts Long sees related to the very new and early changes brought about by computers. He raises concerns that imagination will be lost as humanity increases its involvement with these new technologies of knowledge as opposed to past engagement with technologies of power (those that amplify human power, such as weapons, manufacturing, and transportation).[8] In the face of technologies of knowledge, Christian paradigms may also need to be rethought. To "wrestle from a position of informed concern" as Christians, we need to refine, retrieve, and revisit the best of our social, religious, and scientific resources.[9]

The changes to social practices, relationships, and identities brought about by digital technology indeed require that we deploy moral approaches that are as responsive and flexible as the networking model upon which the technology is built. The process of moral response is about questioning how current social and historical meanings inform our response to "another" and the way our response changes how we understand ourselves. In the action of responding (whether that is proactive or reactive), we create new meaning, sometimes, reinforcing past social and historical meaning or adapting and changing it. Moral growth is about identifying weaknesses and vulnerabilities, sinfulness, or moral malformation; it is a hacking of the self and the social worlds of which we are a part.

We know what physical resources it takes to repurpose a gun into a garden tool. The digital ecosystem is not a material thing to be melted down and reconstructed. Instead, we disrupt and repurpose the ways in which it is used, reshaping this mutable technology and the ecosystem upon which it functions. I argue for four moral approaches and commitments in response to specific digital technological shifts. Of course, there are others, and to discover them, we need to continue to foster moral creativity within communities. The four that I suggested in the earlier chapters are expand diversity, cultivate attunement, practice metanoia, and responsibly cocreate. These four ethical responses are not new in a digital age; however, they are repurposed to fit new circumstances. Even as we consider ways to hack digital technologies, we are also hacking ethics.

The four moral approaches are integral to a responsive process of moral formation and growth. *Diversity* is a hallmark of the created world and what is called forth in a morally enriching experience with the other. The wider the gap of differences between me and another, the more difficult the moral encounter and, perhaps, the more rewarding. *Attunement* is an awareness of and response to feelings, needs, and desires. Attunement brings with it a deep connection to our embodiedness. We respond to another with the fullness of our existence, which includes physical and emotional responses. Cultivating attunement is not about following rules or restraining our baser desires. Attunement is an ongoing relational process similar to descriptions above of creative moral response. The concept of moral growth requires *metanoia*—accountability to past actions, awareness of systemic forces that shape morality, and transformation of personal and communal relationships. More than stewardship of relationships or resources, *responsible cocreation* is an acknowledgment of the interdependence and power of all creation.

Each of the moral approaches addresses a particular issue of inequality or moral deformation perpetuated by digital technologies. Attention to differences as a valued part of creation is necessary when algorithmic bias perpetuates racism. Attunement pushes against reducing humans to metrics or data and sees them instead as relational, embodied, and spiritual beings. Metanoia responds to the surveillance modes used to control people and limit the ways in which they grow and change. Responsible cocreation calls our attention to the ecological degradation brought about by digital technologies and the need to stop exploiting the earth and vulnerable populations globally.

Unfortunately, the majority of us approach digital technologies as something we consume or that helps us consume—information, goods, and services, such as entertainment on demand. These four moral approaches and the needs to which they respond require more of us. Communications theorist Andrew V. Edwards says that users need to "recalibrate yourself from consumer of digital to creator of digital."[10] This reinforces the role of the user as cocreational with God, the earth, and digital technologies. We start with the moral approaches outlined thus far and continue to develop new digital literacies as we hack our way to a just digital society.

Critical awareness implies learning how systems work, developing a socially minded curiosity that often starts with observations about personal experience and connects to larger social patterns. The moral approaches to expand diversity, cultivate attunement, practice metanoia, and responsibly cocreate are part of the communal practice of conscientization. For example, when we investigate how impression management morally forms us as we engage digitally social selves, we reflect on social network behaviors, ours and others, as they are shaped by and respond to systemic forces, such as network design, profit motivation, and use of personal information. Knowing how these built-in structures restrict or promote certain expressions of self and relationship building enables critical moral response. By engaging in this type of reflection on our actions, we increase our digital literacy and capacity to morally respond to new aspects of digital technological change.

Changing the technology from a gun to a garden tool reorients how we make meaning and who we become as we engage this new tool. Though we may not radically restructure the infrastructure of digital technology, we can reorient the purpose of the technologies and our use of them. We need digital literacies to intentionally and successfully accomplish new forms of engagement. If moral growth is what we seek, then we need to keep working in ways that advance diversity, emotional awareness, transformation, and interdependence.

## DIGITAL LITERACIES FOR HACKING

Digital literacies encompass digital skills as well as consideration of the contextual and relational qualities and functions of digital technologies. All technologies, whether digital or not, mediate different kinds of *actions*, *meanings*, *relationships*, *ways of thinking*, and *identities*.[11] Becoming digitally literate allows a user to become a creator. Some of us might become very adept creators, and others of us might tinker around the edges. Either way, to have a moral impact, we cannot simply disconnect nor can we allow technologies and those who create them to be the only designers. As media theorists Rodney Jones and Christoph Hafner argue, "the more we know about how these media work, the better we can become at hacking them through selective appropriation, adaptation, modification and mixing to fit our own purposes and promote our own agendas."[12]

Hacking is often given a negative connotation. However, in software design, hacking is an important feature of improving systems. Hackers identify vulnerabilities in systems. The awareness of these vulnerabilities allows programmers to fix them and create safer, stronger systems. Of course, some hackers exploit vulnerabilities for social and personal gain in ways that harm others. Yet the initial act of identifying the vulnerabilities is still helpful. Hacking is a transformative act that requires facility with digital literacies.

Digital literacies help us identify the affordances (benefits) and constraints (limitations) of digital technologies. Since our experience of the world is mediated through cultural technologies, how we experience it "will always be affected by the affordances and constraints of these tools."[13] For example, digital communication allows us to instantaneously connect across wide distances of geography and time (affordance), but it also leads to us being constantly connected and reachable, thus altering privacy (constraint). Hacking allows us to exploit, change, and redesign these affordances and constraints. Of course, that means we are always developing new versions of technologies, much like the gun becomes a garden tool. Each technology has its own affordances and constraints.

Development of digital literacies to enable hacking is a socio-moral process. Jones and Hafner write, "While a big part of [cultivating] digital literacies is understanding the affordances and constraints of the technological tools available to us, another big part is understanding ourselves, our particular circumstances, our needs, our limitations, and our capacity for creativity."[14] We develop literacies to be able to do things in the world, to "become certain kinds of people, and create certain kinds of societies."[15] With hacking, these literacies can be put in service of harmful or helpful outcomes; usually, it is a mix of both. I am arguing that the activity itself is a social good because of what it shows us about who we have become and what kinds of societies we are creating. As Jones and Hafner conclude, "Perhaps the most important digital literacy you can [cultivate] is learning how to ask the right questions."[16] Asking questions is at the core of moral growth.

Hacking in the most concrete sense of the word is about gaining access to a computer system without permission and exploiting its vulnerabilities. I am scaling up the concept and relating it to an ethical call to gain access to the ecosystem of digital technologies and defining the vulnerabilities as the perpetuation of social inequalities and injustices. For some, this moves hacking from a general action to an activist commitment sometimes referred to as hacktivism. Hacktivism is "politically motivated computer hacking," a form of activism that often involves electronic civil disobedience.[17] It is an automation of activism sometimes relying on people power (number of users) or on computational power (a separate software program) to occupy virtual space. Hacktivism involves alteration or destruction of virtual space in the same way that occupying a park or shutting down traffic functions in physical protests. Denial-of-service attacks stop another entity from getting its message out—such as flooding a server to shut down a website. In other circumstances, hacktivists might want to share their own message—overlaying a website with their own. Regardless of tactic, hacktivist activities are guided by ethical commitments that make this type of hacking a "form of social change."[18]

Many hacktivist groups focus their attention on economic inequalities, censorship, and net neutrality issues. For example, the Electrohippies Collective shut down the World Trade Organization website during the 1999 meeting in Seattle.[19] As recently as 2017, hackers across Twitter pledged to hack the Federal Communications Commission if they did away with net neutrality—the practice that internet service providers and the government must treat all data the same. Net neutrality is a key feature of an open internet. Given that hacktivists are often in support of free speech and reducing economic disparity, net neutrality is a primary social justice issue for them.[20]

Hacktivism is one manifestation of hacking that has social change as its aim. Of course, hacking exists on a spectrum related both to tactics and the ethical norms that guide it. Increasingly, do-it-yourself (DIY) computing is accessible to more and more users. There are also ways to participate in online activism that do not fit the traditional definition of hacking in a computational sense. Users are hacking if we consider what they do from a social perspective related to the disruption of current normative systems of social practice, identities, or relationships.

Twitter is used by many groups primarily for activist agendas, to inform a wide public and include a specific call to action (e.g., #JusticeforTrayvon sought to bring people together for rallies and vigils).[21] Twitter activists also successfully pressured one of the jurors who acquitted George Zimmerman to drop her book contract. Feminista Jones in a column titled "Is Twitter the Underground Railroad of Activism?" writes, "particularly for people of color, Twitter has become one of the most important tools of modern sociopolitical activism, a powerful force in the Zimmerman trial aftermath and beyond."[22]

Some critics point out that proportionally more people of color than Whites in the United States use Twitter.[23] Access to Twitter and the ability to generate large viral networks impact how a story is told or what action is taken. Many refer to this ability as a democratization of media or technology. While many have a legitimate criticism that Twitter allows the privileged to take on someone else's cause in an underinformed, do-gooder manner, activist scholars such as Andrea Smith point out that the privileged (majority stakeholders in the status quo) dismiss Twitter activism as a way to silence minority voices that are raising awareness and organizing via the platform.[24] Activists such as Smith ask us to question whether the content, the mode, or the group is what aggravates critics of Twitter activism. It is probably a combination of the three. It may also be that raising awareness is itself enough of an activist contribution to lead folks over time (a much slower timeline than Twitter uses) to work for concrete change related to a particular issue.

Twitter is not simply a tool for organizing or a space to share random thoughts. Like any significant technology, it alters our behavior and even our thought patterns. How one seeks to organize across the globe has been changed by Twitter. And Twitter has been changed by its users. It adopted the hashtag because of user adaptation of the tagging system. While the creation of key phrases or aphoristic, value-laden slogans has been around for centuries, the spread of such networks and formation of communities reaches new levels with instant sorting and searching. Given the speed and possible anonymity, great harm can be done. There are plenty of reported examples of harm, such as the onslaught of tweets that women of color receive when they speak up against sexual violence. Twitter—like any other network of people, form of communication, and open space for the public—is not free from the same oppressive forces that shape everyday lives. Can it be used to reinforce them or counter them? Yes, as the examples have shown.

Similar to the canon of Scripture and those writings left out, Twitter might offer a view to the values and movements that various communities and even global connections support today. As Andrea Grimes writes, "Activist hashtags, while they may be fleeting, and they may require a bare minimum of engagement from many, also act as memory markers, identifiers and names for mercurial moments and movements that shape our present and our future, but which might otherwise be obscured by the passage of time, as so often happens to our work when it happens online. This is not a failure of online activism itself, but a failure of humans to find good ways to achieve it."[25]

One of the most effective adaptations and exploitations of Twitter's affordances (benefits) can be seen in the organizing of the #BlackLivesMatter movement. Online activism need not hack platforms in a traditional sense of disrupting their service. Hacking can use digital technologies to disrupt social systems and ideologies. Employing digital literacies to hack has a creative dimension that relies on participatory cultures of internet users. Participation is an essential aspect of digital citizenship—moving from consumer to creator of digital technologies. This can be seen in each of the activist examples provided.

## PARTICIPATORY CULTURE AND DIGITAL CITIZENSHIP

Digital citizenship is often defined by a list of rights and responsibilities of users, much like the description of ethics I argue we need to revise. Digital citizenship, like ethics, might be more impactful if we consider it from a perspective similar to moral growth and response. Generally, digital citizenship sees digital literacies as skills to be learned to help citizens function in normatively governed ways that benefit the current structure of the internet. However, this type of participation is not about cultural production and creativity as much as it is about cultural *re*production. A list of rights and responsibilities will not alter the current configurations of unethical digital practices, such as algorithmic bias, metric-based impression management, exploitative dataveillance, or built-in obsolescence of digital devices.

In *Negotiating Digital Citizenship: Control, Contest and Culture*, editors Anthony McCosker, Sonja Vivienne, and Amelia Johns suggest that digital citizenship needs a more "fluid interface" that focuses on the "promises of creative culture and alternative modes of participation."[26] This would mean a move away from approaches related to control toward activities that view the internet and digital technology as a site of cultural contest that enables cyberactivism and political strategizing.[27] A step further would be to encourage digital citizenship as a DIY digital culture. This is about more than expanding participation; it supports "experiment with alternative forms of community, representation, deliberation and understanding of moral obligations and rights of the ethical citizen."[28] This type of digital living is one that embraces the creative moral responsiveness I argue for throughout this book. As noted in the introduction, ethicist John Wall writes that moral response "creates received historical and social meanings into new worlds of meaning over time and in response to others. It deals in moral tension and disruption as selves confront their own narrative diversity and the otherness of others."[29] The similarity of this description of moral responsiveness and digital citizenship is grounded in creativity, relationality, and responsiveness.

Citizenship denotes a citizen and a collective society to which that citizen belongs. Similarly, moral growth happens in an individual, though it is always in response to another or others. Bringing together digital literacies and the ethical approaches outlined above is a description of participatory culture online. Henry Jenkins describes it as "phenomena wherein the confluence of technological accessibility and creative capacity gives individuals the ability to actively create, modify, and participate in the culture of which they were previously just passive consumers."[30] Jenkins and media theorists Mizuko Ito and danah boyd suggest that participatory cultures are "'doing it together' in addition to 'doing it yourself.'"[31] There is commitment as well as an ebb and flow between the individual and the collective. When it comes to inequalities caused by diversities in the collective, "participatory culture has a politicized vision of the world in which more people have access to the means of knowledge and cultural production and have a voice in governance and collective action."[32] Participatory cultures use digital resources, such as online spaces and free, open social platforms, to expand opportunities for participation. Participation itself becomes the good that the group seeks, knowing that creative access offers opportunities for encounter.

Hacking is at the intersection of digital citizenship, digital literacies, and participatory culture. Hacking in the context of participatory culture is not solely, though often includes, alteration of software and platforms. Additionally, hacking creates shifts in culture, such as ideas, identity, and relationships. I find it difficult to draw a stark distinction between alterations to technology and culture because they are deeply intertwined in a digital world. Hacking requires that we employ digital literacies that help us with both user skills and forms of recognition that build, shape, and alter culture.

## CREATIVITY, HACKING, AND COMMUNITY

From makerspaces to fandom sites to platform cooperativism movements, the internet is a space of participatory cultures. These communities are examples of collectives of people who often exhibit a high level of digital skill. They share the use of creative technological adaptations and repurposing to achieve their ends of greater participation and commitment. Alternatively, we will explore what I call hacking lite, or what we might think of as the beginning stages of participatory cultural formation. The purpose of this is to identify concrete and often haphazard ways that digital citizens engage in the processes described throughout the chapter. Also, because the digital and physical worlds are identifiable yet not distinct spaces in our lives, I turn to an example of participatory culture that is both online and offline.

The Episcopal School of Los Angeles (ESLA) is a community that models creative engagement in relation to both moral formation and digital technologies, bringing together the concepts in this chapter related to an ethic of hacking. The ESLA is a participatory culture in the best sense of the term—increasing technological accessibility, developing creative capacities, and widening participation. The ESLA integrates the four ethical approaches I have outlined related to an ethic of hacking—fostering diversity, cultivating attunement, encouraging metanoia, and cocreating responsibly. The ESLA lives out this vision in an aspirational sense related to its presence as a hybrid digital and enfleshed community. A variety of participatory communities exist online and offline, with each space influencing the other.[33] The ESLA is a helpful example because it is explicit about its value framework, educational goals, and creative and informed engagement with digital technologies. Given the integrated nature of a learning community, it is difficult to pull out individual examples related to the ethical approaches. Instead, through a robust description of the ESLA community, I will highlight ways their participatory community evidences overlapping and unique practices.

The ESLA is an independent sixth- through twelfth-grade school started in 2012 that sprang up out of an after-school STEM (science, technology, engineering, and math) program in 2009 and graduated its first high school senior class in 2017.[34] It is a liberal arts school that holds to its roots of STEM education by promoting various scientific literacies. As this relates to digital technology, the ESLA does not separate a digital or online world from an in-person world. For example, the living out of school values is applied across all spaces. Whether a student does something in person or on Facebook is irrelevant; the conduct is what matters, not the location. There is no formal system of surveillance to which the ESLA subscribes. Instead, the community focuses on shared values, shared governance, and accountability to values across online and offline spaces mediated through varied and overlapping networks. Its approach fosters moral growth by encouraging metanoia, or social and personal transformation.

The ESLA has core values that stem from its mission statement—generosity, curiosity, and courage. When there is a question about or incident where a student has potentially violated these values, a student- and faculty-led committee performs the review. Students and faculty collectively wrote of the honor code that guides the review. These processes engage the moral formation of students and thus the full community by giving the members a role in articulating the meaning of core values and how they show up (or don't) in interactions with one another. Additionally, all classes share learning outcomes related to joy, community, dialectical thinking, and diversity. For example, many classes start with a common writing prompt whereby a faculty member writes a statement on the board and students free write a defense of the statement. Then, a few minutes later, the teacher reverses the statement, and students must defend the opposing argument. This process is about cultivating dialectical thinking, but it also requires generosity, curiosity, and courage (especially if they are shared with others). Having the skill and flexibility to take opposing points of view opens the students to seeing issues from a variety of perspectives and listening with care to one another. The missional, or core, values and learning outcomes reciprocally build on each other.

When it comes to digital technologies specifically, the ESLA has a variety of approaches. The use of personal digital devices is not allowed during the school day. This is one way to model boundaries and encourage attunement to personal platforms that perpetuate the need to constantly craft and curate one's self. Each student is provided with a Chromebook, thus creating a shared digital ecosystem for use of digital technology in the classroom. However, these uses are specific to the classroom context and learning relationships and focus on literacies related to information, research, creation, and adaptation of various media. The shared ecosystem sets up perimeters related to privacy of information and surveillance of online participation as they relate to schoolwork and research. The use of a shared system balances the need for closure and openness in learning environments without perpetuating the overaccumulation of data and static moral development that some forms of dataveillance or social surveillance can cause. Students and teachers alike are expected to learn, change, and adapt as they engage new technologies of learning—some digital, some analog.

The shared digital ecosystem is not necessarily bound by the walls of the school. Students may walk to a nearby art gallery and use their Chromebooks for shared projects. In 2014, students who traveled to Costa Rica were challenged to leave behind their social media connections and use their hands as "advanced technology." During that process, however, they did not leave all digital technology behind. They used sensing devices to connect to the sounds of the rainforest; wrote for a class blog that informed parents and the school community about what they were learning; and ultimately, captured the trip in a video format shared on YouTube.[35] They used digital technologies in service of connecting to creation; deepening relationships with each other and family across geographic distance and remembering in a new form create an ongoing experience and memory of the trip.

While some practices at the ESLA rely on physically connecting to the created world, others are about digital cocreation. The ESLA also has a nascent media lab, built in the same vein as the MIT Media Lab, where a cross pollination of people come together in an open working space. In fact, there is not enough room for an analog library space, so the Media Lab serves the purpose of electronically connecting students to the analog books, magazines, and texts they need for schoolwork. What the ESLA has found is that while many students and adults are technology dependent, they are not necessarily technologically savvy or "naturally" inquisitive. The ESLA hired a full-time director for the media lab, or Incubator, to resource the community and encourage play and investigation. They realized that there is and will be a significant amount to be taught related to how, why, and in what ways to use digital technologies, such as a 3D printer or robotics equipment.

As the ESLA literature states, "we want our students to be able to ask the larger questions about how technology shapes the ethical land- scape of the world they live in, and how it can be harnessed toward positive social transformations."[36] Oftentimes, teachers and administra- tors model this without having to "teach it" in an explicit sense. Provid- ing everyone with the same technology in the classroom equalizes ac- cess and provides a common platform from which to engage. The school also ensures that students have a reliable internet connection when they are home. These are a few ways that the ESLA promotes technological accessibility that pays attention to physical and virtual disparities to increase diversity.

The ESLA, like any community, could do more. As a newly founded school, it has begun from a place of shared values that are challenged and adapted as it grows, inviting creative moral growth. It even fosters what I have called attunement related to digital self-formation. Rev. Megan Hollaway in a course on the Hebrew Scriptures assigned sixth- graders a digital Sabbath for a full weekend as a way of understanding the concept of Sabbath and practicing a habitual disengagement from social networking. The assignment, not loved by all, brought up ques- tions about the affordances and constraints of social networking on the everyday experiences of these youth. Faculty members also commit to having their digital devices stowed in a pocket or bag while in common areas or walking between spaces to promote and model face-to-face engagement. It calls forth questions of how digital lives are currently shaped, how we want them to be shaped, and what types of community social networks create. Through its policies and practices, the ESLA cultivates attunement by setting boundaries on the use of personal and school devices as well as expectations about behavior across social net- works. As the faculty, administrators, and students engage digital tech- nologies, they do so with an ethical intentionality that they bring to all learning and living.

One of the assets and commitments of the ESLA is diversity, which is expanded in its curricular engagement with social justice and direct service to the wider community of Los Angeles. It provides a significant amount of financial aid to create a diverse student body across racial, ethnic, and socioeconomic divides in Los Angeles. Participation and access to participation are key features of the ESLA. The ESLA, while an Episcopal school, does not ask students or teachers to espouse a particular doctrine. The ESLA "would rather provide our students with the tools to examine and articulate their own beliefs. The Episcopal tradition teaches us that it is far better to live in the tension of difficult questions than in the static certainty of answers."[37] The intersection of diversity and the Episcopal identity of the ESLA comes together in the service experiences that students and faculty share. The ESLA reflects the city of Los Angeles and engages the city. Students serve in soup kitchens and homeless shelters. They use the city as a classroom in the best sense of the word, not shying away from economic, religious, racial, or political differences, as students become informed and responsible citizens.

The hybrid participatory culture of the ESLA empowers students to be hackers not only to adapt digital technologies but also to create new cultures of accountability that are diverse; aware of social dynamics on personal, communal, and systemic levels; and grounded in moral growth. As mentioned earlier, "Perhaps the most important digital literacy you can [cultivate] is learning how to ask the right questions."[38] Hacking is a form of finding new possibilities based on questions. The framework within which the ESLA does this reflects the four moral approaches detailed previously. In both online and offline spaces, the ESLA seeks to increase diversity; promote practices of attunement; teach and govern in ways that encourage metanoia; and provide experiences that expand responsibility as cocreators to community formation, digital technologies, and the intersection of both.

## CONCLUSION

The metaphor of beating swords into plowshares reminds us that technologies can have a variety of purposes depending on how they are constructed and used. Humans are involved at all stages of these processes. While not every user of digital technology needs to know or even could know the intricacies of the digital ecosystem, everyone can benefit from developing or deepening their digital literacies. Christians who work toward social change, such as reducing economic inequality, ending racism, and responding to ecological degradation, need look no further than their digital devices. That does not mean digital technology only perpetuates these injustices. In fact, many Christians use them for creative means to respond to these injustices, including participating in virtual activist campaigns, learning new information, or connecting like-minded people with collaborators.

When our engagement with digital technologies stays on the level of consumption, we are more susceptible to promoting corporate interests rather than Christian ethical commitments. Thinking theologically with and about digital technology pushes us to find language within the Christian tradition to respond to new challenges and circumstances. I began this book by stating that God is with us, calling us to moral response. Throughout, I reiterate a commitment to seeking out and valuing difference in line with God's intentions for creation. We find God at the Tower of Babel affirming difference in response to the use of sameness for empire building. In digital technology discussions, some scholars focus on the technology of building at Babel and assume that God's reaction suggests human technology in the form of building is bad. When we focus on the use of language in the text, God's response is not against technology as building; rather, it is against the reduction of cultural difference through the limiting of language. Gaining technological literacy is yet another language, and digital literacy is an avenue to expand social and cultural difference in our networks.

The shaping of our social networks via digital technology directly affects our sense of self and either expands or reduces differences we encounter. As embodied and spiritual beings, Christians wrestle with how much digital profiles, trails, and transactions shape us. Data makes up more and more of who we are and how we are in the world. Trinitarian understandings of God model the interdependence and intimacy needed to theologically imagine new forms of self as datafied, embodied, and spiritual. Additionally, moral practices of attunement, forgiveness, and metanoia provide grounding in a rapidly shifting digital environment. Attunement as a moral practice or way of being relies on an embodied and relational awareness of self and responsiveness to the other and God. The level of data and social surveillance in our everyday lives appears to trap us in pasts that are part of an eternal memory; we can lean into Christian acts of forgiveness and metanoia to transform ourselves, our relationships, and social structures. Metanoia does not rely on digital practices of erasure; rather, it nurtures Christian practices of accountability.

The interdependence of God as creator, sustainer, and redeemer and difference as a core part of creation converge in an expansive biodiversity that brings the earth and all that comprises it into focus. Earth is intricately woven into digital devices as we uncover how they are made of organic material from around the world and follow the vibrational effects of networks of fiber optic cables and server farms that power the digital network. The digital ecosystem is interdependent and transforming earth's environment, which requires that we deepen our ability to sense creation and matter in digital connections in material form and in moral value.

Creative adaptation, repurposing, and modification are ethical ways that we engage digital technologies when we adapt the moral approaches of expanding diversity, cultivating attunement, practicing metanoia, and cocreating responsibly. When we intentionally promote diversity in our networks by changing our settings or joining the Algorithmic Justice League project, we are exposing the vulnerabilities of the current system. We may not do this at the level of hacktivists who shut down systems to draw attention to justice issues, such as net neutrality, but a variety of actions are needed. Hacking may seem like a distant concept when one is just beginning to hone one's digital literacies. But hacking need not be practiced only by those who know how to program software or build digital devices. Some will work on systemic levels and others on individual levels.

The fluid and malleable nature of digital technologies calls for creativity as moral response. The community practices at the ESLA and at RAWtools demonstrate how ethical consideration of the role of technologies can lead to imaginative and resourceful outcomes. Individually, we may start with simple hacks to modify use, such as turning off the notification lights on our mobile devices to disrupt the desire-shaping forces of social media. Or we might reassess the value of our data to cut down on the free capital that we not only share but that permanently marks us with digital trails. We could audit our networks for diversity and intentionally seek it out. For those with more software skills, we can join movements to alter platforms for cooperative sharing, not just free enterprise. All of us need to go back to school in a certain sense. Like the ESLA, we must assess how we collectively share in and take responsibility for technology use, access, and development as we seek to form the moral values of our communities.

Each of us can intentionally reconnect to the amount of energy we consume to power digital technologies. We might find out where our devices come from or how they were made. Digital devices are meant to seem like an extension of our bodies and minds. Digital has changed the way we are as human beings—datafied and networked selves, shifting social identities, relationships, and practices. Moral growth as a creative response to the other includes our response to digital technologies and provides an opportunity to renew our theological foundations.

Wonder, imagination, and creativity are the genesis of digital technology; they also must be its constant moral companion. As Christians, we need to learn about how the digital ecosystem functions, from the mining of materials to building devices to the impact of algorithms. The process of demystifying and desacralizing technology empowers Christians to morally evaluate all aspects of digital growth, adaptability, and responsiveness. That process requires developing digital literacies and optimally takes place in the context of a participatory community whether that is a classroom or our family room, a church, or a chatroom. Digital technologies are social; thus, our ethical responses have to account not just for the internal sociality of the technologies but also for their deeply social impact. These shifts do not need to happen *to* us; we must proactively participate in them.

## EXCURSUS 5: READING AND WRITING NEW VISIONS

Then the Lord answered me and said:
Write the vision;
make it plain on tablets,
so that a runner may read it.
For there is still a vision for the appointed time;
it speaks of the end, and does not lie.
If it seems to tarry, wait for it;
it will surely come, it will not delay.

—Habakkuk 2:2–3, NRSV

I have long wanted to be a computer hacker. I am fascinated by the level of knowledge, playfulness, and cooperation present in hacker communities. Of course, not all hackers use their talents for good ends, and there may be a bit of a twisted Robin Hood quality to much of the hacker ethos. Taking systems apart and finding their vulnerabilities with the intention of making them stronger is the aspect of hacking to which I am drawn. After numerous failed attempts at learning to code, I find I never have the time or the will to really learn the craft of computer hacking.

Habakkuk reminds me a bit of an ancient hacker tasked with the role of highlighting vulnerabilities in the current system. Habakkuk's prophesy represents a "complaint against the prosperity of the wicked"[39] by providing "a trenchant critique of imperial brutality."[40] The passage quoted above is bookended by indictments, complaints, and woe oracles—Habakkuk knows the system well. The prophet not only criticizes the wrongdoing happening in Judah but also questions God's plan to correct the corruption by sending the Babylonians, who in the prophet's view, will only create greater violence.

Habakkuk is told to "write the vision," and if it doesn't come as quickly as Habakkuk would like, he should "wait for it." Biblical scholar Theodore Hiebert notes, "Even in the very events and agents understood by Habakkuk as instruments by which divine justice will be done, injustice seems to be present. Real-world politics appear to be continually at odds with the prophetic passion for justice and faith in God's just rule."[41] Habakkuk turns out to be much more patient than most folks I know, including myself, who want a clear and quick response from God about how to address the current, unjust state of affairs. Hiebert continues, "The challenge of believing in the ultimate power of justice in a world that appears to be overwhelmingly unjust is one of the most difficult existential struggles the religious person must face."[42]

The Habakkuk text shows that those who help us find our way toward God are not afraid to ask difficult questions, to critique the social system as it currently exists, and to believe in the possibility of change, even when they don't know or sometimes doubt how it will come to fruition. They are aware of God with them and for them, and they take seriously their calling as cocreators and participants with God. I'm not sure I would want to do this work on my own like Habakkuk did. I much prefer the community provided by a classroom, church, or even friends as we struggle through how to be the people God calls us to be in this current digital society.

What needs hacking in your life or community? How can systems of injustice be disrupted or hacked so we can remake them in new and stronger ways?

# NOTES

## INTRODUCTION

1. Samuel Arbesman, *Overcomplicated: Technology at the Limits of Comprehension* (New York: Current, 2016), 6.

2. The traditions of feminist and womanist ethics have spent decades debunking the myth of a rational and objective, usually male model of decision making, from the fields of psychology to religious studies. However, even when it is counter to everyday experience, I find that students in my ethics classes continue to hold onto the notion of moral decision making as a mix between following set rules and rationally thinking through a problem like a calculator that will spit out the correct answer.

3. I do this in conversation with a variety of scholars, most notably Cristina L. H. Traina, Danna Nolan Fewell, John Wall, Jack Halberstam, Joyce Ann Mercer, Almeda Wright, Jennifer Beste, and Jennifer Harvey, who come from a variety of disciplinary starting points but share the commitment to children and youth as capable and astute, fully moral people. I address these issues in a number of my own writings, specifically "Animating Children," symposium on *The Queer Art of Failure* by Judith (Jack) Halberstam (Durham, NC: Duke University Press, 2011), June 2015, https://syndicate.network/symposia/theology/the-queer-art-of-failure; and "Children as An/other Subject: Redefining Moral Agency in a Postcolonial Context," *Journal of Childhood and Religion* 5, no. 2 (May 2014).

4. John Wall, *Ethics in Light of Childhood* (Washington, DC: Georgetown University Press, 2010), 169.

5. Wall, *Ethics in Light of Childhood*, 169.

6. Wall, *Ethics in Light of Childhood*, 169.

7. Wall, *Ethics in Light of Childhood*, 176.

8. Wall, *Ethics in Light of Childhood*, 171.

9. Emilie M. Townes, *Womanist Ethics and the Cultural Production of Evil* (New York: Palgrave Macmillan, 2006), 164.

10. Wall, *Ethics in Light of Childhood*, 177.

11. Richard M. Gula, *Just Ministry: Professional Ethics for Pastoral Ministers* (Mahwah, NJ: Paulist Press, 2010), 33. Gula is referencing a remixed version of the title of the book *Earth Crammed with Heaven: A Spirituality of Everyday Life* by Elizabeth Dreyer (Mahwah, NJ: Paulist Press, 1994).

12. Paulo Freire, *Pedagogy of the Oppressed*, 30th anniversary ed., trans. Myra Bergman Ramos (New York: Continuum, 2005), 126.

13. Rodney H. Jones and Christoph A. Hafner, *Understanding Digital Literacies: A Practical Introduction* (New York: Routledge, 2012), 13.

14. John Hartley, *The Uses of Digital Literacy* (New Brunswick, NJ: Transaction Publishers, 2012), 21.

15. Andrew Burn, *Making New Media: Creative Production and Digital Literacies* (New York: Peter Lang Publishing, 2009), 22.

16. Hartley, *Uses of Digital Literacy*, 23–28.

17. Jones and Hafner, *Understanding Digital Literacies*, 5–9.

18. Jones and Hafner, *Understanding Digital Literacies*, 12.

19. Yoram Eshet-Alkalai and Eran Chajut, "Rapid Communication: Changes over Time in Digital Literacy," *Cyberpsychology & Behavior* 12, no. 6 (2009): 713–15.

20. Jones and Hafner, *Understanding Digital Literacies*, 111.

21. Lee Rainie, "Digital Divides—Feeding America," Pew Research Center, accessed September 20, 2017, http://www.pewinternet.org/2017/02/09/digital-divides-feeding-america.

22. Some examples include John Dyer, *From the Garden to the City: The Redeeming and Corrupting Power of Technology* (Grand Rapids, MI: Kregel Publications, 2011); Leonard Sweet, *Viral: How Social Networking Is Poised to Ignite Revival* (Colorado Springs, CO: WaterBrook Press, 2012); Shane Hipps, *Flickering Pixels: How Technology Shapes Your Faith* (Grand Rapids, MI: Zondervan, 2009); Paul Asay, *Burning Bush 2.0: How Pop Culture Replaced the Prophet* (Nashville: Abingdon Press, 2015); and Craig Detweiler, *iGods: How Technology Shapes Our Spiritual and Social Lives* (Ada, MI: Brazos Press, 2013). This is not to say that there are no women or people of color writing on the subject; one example is Elizabeth Drescher, *Tweet If You [Heart] Jesus: Practicing Church in the Digital Reformation* (Harrisburg, PA: Morehouse Publishing, 2011).

23. I am in conversation throughout the coming chapters with many of these authors. When possible, I identify their professional and scholarly loca-

tion so readers are aware of the multidisciplinary conversation and theologically diverse conversation partners.

## 1. PROGRAMMING FOR DIFFERENCE

1. Michael Feldman, Sorelle Friedler, John Moeller, Carlos Scheidegger, and Suresh Venkatasubramanian, "Certifying and Removing Disparate Impact," *Machine Learning*, July 16, 2015 (version 3), accessed June 16, 2017, https://arxiv.org/pdf/1412.3756.pdf.

2. Eli Pariser, *The Filter Bubble: What the Internet Is Hiding from You* (New York: Penguin, 2011), 9.

3. See Rodney H. Jones and Christoph A. Hafner, *Understanding Digital Literacies: A Practical Introduction* (New York: Routledge, 2012), chapter 2, "Information Everywhere," especially pages 19–22.

4. This is not to say that language has not been used for similar purposes in the past; however, in the past, humans are the subjects using language.

5. In Christian readings of the text, Babel is often linked to Pentecost (Acts 2:1–42). Some Christians assert that Pentecost is a correction, or overturning of the sin of Babel. If Babel is about God punishing people with multiple languages, Pentecost is read as God erasing the problem of multiple languages so that the disciples' preaching and evangelizing is heard by all. In other words, this reading suggests that through the Spirit's infusion, early Christians at Pentecost got "sameness" right, unlike the builders at Babel. See José Míguez-Bonino, "Genesis 11:1–9, A Latin American Perspective," in *Return to Babel: Global Perspectives on the Bible*, ed. John R. Levison and Priscilla Pope-Levison (Louisville, KY: Westminster John Knox, 1999), 13–16. In Míguez-Bonino's interpretation, upon which I will rely, God employs a form of technology—language—to encourage the building of something much richer and more complex. Pentecost then serves as one example of how unity in difference requires understanding rather than erasure.

Míguez-Bonino notes that the Pentecost story, as separate from any relationship to Babel, raises a distinct question about the relationship between the universal claims of Christian faith and the particularity of peoples. He writes, "what the 'gospel of Pentecost' tells us clearly is that the 'universality' of the history of salvation is not the dissolution of specific spaces—ethnic, cultural, linguistic—which are God's own creation and which are constantly being renewed by God's Spirit. Rather the 'universality' of the history of salvation is the negation of spaces closed in themselves and the affirmation of a space open to 'the other' as a revelation of the same universal grace" (164–65). Thus, the distinctly Christian message of Pentecost for Christian believers promotes an interpretation of unity that requires affirmation and understanding, not simply accommodation, of differences.

Sharing the many interpretations of Pentecost by other biblical scholars, Míguez-Bonino concludes that it is clear "Luke wants to emphasize the universal reach of God's message" (163). Some suggest that Pentecost established a new language representative of the covenant of Christ. However, Luke clearly states that the diversity of people gathered heard in "their own languages" (Acts 2:8). Pentecost is not a Christian "fix" to the Jewish or Israelite problem of Babel, though I recognize that even the need to say this means that the juxtaposition or linking of a Hebrew scripture and Christian scripture sets up an if/then algorithmic-like relationship that is difficult to disentangle from Christian hegemony. Míguez-Bonino also identifies a possible link with the Jewish tradition "when God gave the law on Sinai, God's word was divided into different languages so that all people could hear it" (163). This would be another scriptural reference to God's active involvement in engineering difference.

6. Before we delve deeply into the biblical stories, I provide a brief interlude about scriptural interpretation. As Christians, we use the scriptures differently across our denominational divides and even within our denominations. Some argue fervently for one way to interpret a text or even the whole book. Others see scripture as a historical and literary document that carries meaning for many people based in their connection to the communities who produced the texts, sometimes inconsistently and contradictory, generally not literally. Believers' interpretation of scripture is much like Pentecost, Peter Gomes has written: "The history of the interpretation of scripture is a history of the capacity of God's word to speak in many tongues and in many ways, and to draw all people into its gracious embrace. It strikes me that that is perhaps the most compelling and exciting case for what we would call the 'authority' of scripture." See Peter Gomes, "Beyond the Human Point of View," presentation at the Covenant Network of Presbyterians Luncheon, 211th General Assembly, Fort Worth, Texas, June 21, 1999, accessed June 6, 2014, http://covnetpres. org/1999/06/beyond-the-human-point-of-view. Biblical stories have power precisely because of the myriad of ways they can and are interpreted, providing meaning and direction for God's people. Even this approach to scriptural interpretation embraces God's intended diversity and difference.

7. Míguez-Bonino, "Genesis 11:1–9," 13–16. See also J. Serverino Croatto, "A Reading of the Story of the Tower of Babel from the Perspective of Non-Identity Gen 11:1–9 in the Context of Its Production," in *Teaching the Bible: The Discourses and Politics of Biblical Pedagogy*, ed. Fernando F. Segovia and Mary Ann Tolbert (Maryknoll, NY: Orbis Press, 1998), 203–23. Croatto provides a more historically contextual reading of the Babel passage and concludes

that "Gen 11:1–9 problematizes the unity of language, from the point of view of human hubris or excess, as an instrument of oppression" (222).

8. Míguez-Bonino, "Genesis 11:1–9," 161–65.

9. Brian Brock, *Christian Ethics in a Technological Age* (Grand Rapids, MI: Eerdmans, 2010).

10. Brock, *Christian Ethics*, 225.

11. Brock, *Christian Ethics*, 227.

12. Brock, *Christian Ethics*, 230.

13. Quentin J. Schultze, *Habits of the High-Tech Heart: Living Virtuously in the Information Age* (Grand Rapids, MI: Baker Academic, 2002), 49.

14. Schultze, *Habits of the High-Tech Heart*, 68.

15. Schultze, *Habits of the High-Tech Heart*, 52–54.

16. Schultze, *Habits of the High-Tech Heart*, 196.

17. John Dyer, *From the Garden to the City: The Redeeming and Corrupting Power of Technology* (Grand Rapids, MI: Kregel Publications, 2011), 103.

18. Dyer, *Garden to the City*, 105.

19. Dyer, *Garden to the City*, 105.

20. Dyer, *Garden to the City*, 105.

21. Míguez-Bonino, "Genesis 11:1–9," 13. Emphasis in original.

22. Míguez-Bonino, "Genesis 11:1–9," 14–15.

23. Míguez-Bonino, "Genesis 11:1–9," 15.

24. Míguez-Bonino, "Genesis 11:1–9," 15.

25. Míguez-Bonino, "Genesis 11:1–9," 15–16. Emphasis in original.

26. Jack Lule, *Globalization & Media: Global Village of Babel* (Lanham, MD: Rowman & Littlefield, 2012), 10.

27. Chava Gourarie, "Investigating the Algorithms That Govern Our Lives," *Columbia Journalism Review*, April 14, 2016, accessed June 15, 2017, https://www.cjr.org/innovations/investigating_algorithms.php.

28. Schultze, *Habits of the High-Tech Heart*, 26.

29. Gourarie, "Investigating the Algorithms."

30. Jessica Guynn, "Google Photos Labeled Black People 'Gorillas,'" *USA Today*, July 1, 2015, accessed September 20, 2017, https://www.usatoday.com/story/tech/2015/07/01/google-apologizes-after-photos-identify-black-people-as-gorillas/29567465.

31. Roberto Simanowski, *Data Love: The Seduction and Betrayal of Digital Technologies* (New York: Columbia University Press, 2016), 50. See also Safiya Umoja Noble, *Algorithms of Oppression: How Search Engines Reinforce Racism* (New York: New York University Press, 2018).

32. Simanowski, *Data Love*, 51.

33. Simanowski, *Data Love*, 56.

34. Simanowski, *Data Love*, xv.

35. Simanowski, *Data Love*, 42.

36. Public Religion Research Institute, "Race, Religion, and Political Affiliation of Americans' Core Social Networks," August 3, 2016, accessed June 6, 2017, http://publicreligion.org/research/2014/08/analysis-social-network.

37. I address this research and issues of homophily in social networks related to feminist communities in Kate Ott, "Hacking the System," *Journal of Feminist Studies in Religion* 31, no. 2 (2015): 140–44.

38. See Saul Levmore and Martha Nussbaum, eds., *The Offensive Internet: Speech, Privacy, and Reputation* (Boston: Harvard University Press, 2010).

39. Henry Jenkins, "Defining Participatory Culture," in *Participatory Culture in a Networked Era: A Conversation on Youth, Learning, Commerce, and Politics*, ed. Henry Jenkins, Mizuko Ito, and dana boyd (Malden, MA: Polity Press, 2006), 3. The book is a conversation of all three authors with individually authored introductions to each chapter. In the introduction, Jenkins compares the use of participatory culture as a descriptor of online communities and social media in particular to his earlier work on fandom.

40. See Jenkins, "Defining Participatory Culture," 11–12.

41. Astra Taylor, *The People's Platform: Taking Back Power and Culture in the Digital Age* (New York: Metropolitan Books, 2014), 18–19, 218.

42. Andrew V. Edwards, *Digital Is Destroying Everything: What the Tech Giants Won't Tell You about How Robots, Big Data, and Algorithms Are Radically Remaking Your Future* (Lanham, MD: Rowman & Littlefield, 2015), 59.

43. See Wages for Facebook, http://www.wagesforfacebook.com and related articles such as E. Alex Jung, "Wages for Facebook," *Dissent*, Spring 2014, accessed June 15, 2014, http://www.dissentmagazine.org/article/wages-for-facebook.

44. Taylor, *People's Platform*, 228–30.

45. Facebook publishes updates to users on its platform. Additionally, see timelines of Facebook News Feed algorithms such as this one published by Wallaroo Media, updated May 11, 2017, accessed June 30, 2017, http://wallaroomedia.com/facebook-newsfeed-algorithm-change-history.

46. Molly Niesen, "Love, Inc.: Toward Structural Intersectional Analysis of Online Dating Sites and Applications," in *The Intersectional Internet: Race, Sex, Class, and Culture Online*, ed. Safiya Umoja Noble and Brendesha M. Tynes (New York: Peter Lang Publishing, 2016), 162.

47. Niesen, "Love, Inc.," 168.

48. Shoshana Amielle Magnet, *When Biometrics Fail* (Chapel Hill: Duke University Press, 2011), 20.

49. Magnet, *When Biometrics Fail*, 151.

50. Sorelle Friedler, "Being Hopeful about Algorithms," *Algorithmic Fairness, Discrimination and Disparate Impact Blog*, February 3, 2017, accessed June 15, 2017, http://fairness.haverford.edu/index.html.

51. Lauren Silverman, "Using Data to Predict Child Abuse," Marketplace, April 14, 2016, accessed December 8, 2016, http://www.marketplace.org/2016/04/14/world/using-data-predict-child-abuse-hot-spots.

52. Darian Woods, "Who Will Seize the Child Abuse Prediction Market?," *The Chronicle of Social Change*, May 28, 2015, accessed December 8, 2016, https://chronicleofsocialchange.org/featured/who-will-seize-the-child-abuse-prediction-market/10861.

53. Keith Kirkpatrick, "Battling Algorithmic Bias," *Communications of the ACM* 59, no. 10 (2016): 16–17, accessed June 29, 2017, https://cacm.acm.org/magazines/2016/10/207759-battling-algorithmic-bias/fulltext.

54. Friedler, "Being Hopeful about Algorithms."

55. Pariser, *The Filter Bubble*, 3, 113.

56. Pariser, *The Filter Bubble*, 88.

57. Edwards, *Digital Is Destroying Everything*, 85.

58. Samuel Arbesman, *Overcomplicated: Technology at the Limits of Comprehension* (New York: Current, 2016), 6–7.

59. Arbesman, *Overcomplicated*, chapter 3.

60. "Critical Algorithm Studies: A Reading List," Social Media Collective, accessed June 26, 2017, https://socialmediacollective.org/reading-lists/critical-algorithm-studies.

61. Simanowski, *Data Love*, chapter 16.

62. Pariser, *The Filter Bubble*, 233.

63. See Virginia Eubanks, *Automating Inequality: How High-Tech Tools Profile, Police, and Punish the Poor* (New York: St. Martin's Press, 2017), for more on the overlap between poverty and algorithmic bias. There is significant overlap between structural digital oppressions related to class and race as well as gender, age, and disability in many contexts throughout the United States. My focus on racial bias in this chapter is one aspect of difference that needs attention; it is not meant to be to the exclusion of other differences that generate oppressive circumstances.

64. Theresa Senft and Safiya Umoja Noble, "Race and Social Media," in *The Social Media Handbook*, ed. Jeremy Hunsinger and Theresa Senft (New York: Routledge, 2014), 107–25.

65. To learn more about the Algorithmic Justice League and join the fight against algorithmic bias, visit https://www.ajlunited.org.

66. Feldman et al., "Certifying and Removing Disparate Impact."

67. Byrd Pinkerton, "He's Brilliant, She's Lovely: Teaching Computers to Be Less Sexist," All Tech Considered, August 12, 2016, accessed August 12,

2016, http://www.npr.org/sections/alltechconsidered/2016/08/12/489507182/
hes-brilliant-shes-lovely-teaching-computers-to-be-less-sexist.

68. J. Philip Wogaman, *Christian Ethics: A Historical Introduction* (Louisville, KY: Westminster John Knox Press, 1993), 276.

## 2. NETWORKED SELVES

1. See Romanus Cessario, OP, *The Moral Virtues and Theological Ethics* (Notre Dame, IN: Notre Dame Press, 2009), 46. In both definitions, we need to define what is "good" or "morally praiseworthy." In some cases, ethicists have leaned toward describing the good as a fixed set of global or universal principles that often come about through some form of objective reasoning. We are then left with a set of rules that supposedly cross cultures and historical time periods. One can know the "good" or "right" choice if one just thinks long and hard about a moral decision. For example, sexting—sending a nude photo to another person via digital technology—could always be considered morally wrong if there were a universal principle against the sharing of naked images of the human body. From our own experience, we know this description of virtue and the good is too narrow to account for the way we live out our values and beliefs. In fact, across cultures, some of the most famous artworks involve nudes. On the other hand, there is also extensive debate about the moral harms of pornography, both still images and film.

2. Jonathan Haidt and Craig Joseph, "The Moral Mind: How Five Sets of Innate Intuitions Guide the Development of Many Culture-Specific Virtues, and Perhaps Even Modules," in *The Innate Mind, Volume 3: Foundations and the Future*, ed. Peter Carruthers, Stephen Laurence, and Stephen Stich (Oxford Scholarship Online, 2008), 386, accessed July 3, 2017, http://www. oxfordscholarship.com/view/10.1093/acprof:oso/9780195332834.001.0001/ acprof-9780195332834-chapter-19.

3. See Haidt and Joseph, "The Moral Mind." Haidt and Joseph propose five innate foundations for virtue cultivation, like taste buds (385) of the moral sense that are later developed into virtues dependent and responsive to individual personality, relationships, and cultural systems. They also discuss how these taste buds differ from early moral psychologists' assertions that children are blank slates or that morality is a set of rules or principles to be learned over time.

4. Haidt and Joseph, "The Moral Mind," 387.

5. Cristina Traina, "Erotic Attunement," in *Professional Sexual Ethics: A Holistic Ministry Approach*, ed. Patricia Jung and Darryl Stephens (Minneapolis, MN: Fortress, 2013), 44. For more on the development of Traina's concept,

see Cristina L. H. Traina, *Erotic Attunement: Parenthood and the Ethics of Sensuality between Unequals* (Chicago: Chicago University Press, 2001).

6. Traina, "Erotic Attunement," 241.

7. Traina, "Erotic Attunement," 217.

8. James F. Keenan, SJ, *Virtues for Ordinary Christians* (Franklin, WI: Sheed & Ward, 1996), 4.

9. Keenan, *Virtues*, 7.

10. Keenan, *Virtues*, see chapter 15.

11. Dwight J. Friesen, *Thy Kingdom Connected: What the Church Can Learn from Facebook, the Internet, and Other Networks* (Grand Rapids, MI: Baker Books, 2009), 26.

12. Friesen, *Thy Kingdom Connected*, 64–65.

13. Friesen, *Thy Kingdom Connected*, 69.

14. Friesen, *Thy Kingdom Connected*, 55–56.

15. Astra Taylor, *The People's Platform: Taking Back Power and Culture in the Digital Age* (New York: Metropolitan Books), 2014.

16. For a discussion related to the history of the terms, see Peter Fischer-Nielsen and Stefan Gelfgren, "Conclusion: Religion in a Digital Age: Future Developments and Research Directions," in *Digital Media, Social Media and Culture: Perspectives, Practices and Futures*, ed. Pauline Hope Cheong, Peter Fisher-Nielsen, Stefan Gelfgren, and Charles Ess (New York: Peter Lang Publishing, 2012), 296–98.

17. I discuss this in greater detail in Kate Ott, "Social Media and Feminist Values: Aligned or Maligned?," *Frontiers: A Journal of Women's Studies* 39, no. 11 (2018): 93–111.

18. "Three Technology Revolutions," Pew Research Center, accessed October 10, 2015, http://www.pewinternet.org/three-technology-revolutions.

19. Pauline Hope Cheong and Charles Ess, "Introduction: Religion 2.0? Relational and Hybridizing Pathways in Religion, Social Media, and Culture," in *Digital Media, Social Media and Culture: Perspectives, Practices and Futures*, ed. Pauline Hope Cheong, Peter Fisher-Nielsen, Stefan Gelfgren, and Charles Ess (New York: Peter Lang Publishing, 2012), 12. Here, digital media is defined as computer-mediated communication affiliated with Web 2.0, such as social media, including social networking sites (e.g., Facebook), blogs and microblogs (e.g., Twitter), user-generated content (e.g., YouTube), and virtual worlds/games (e.g., Second Life). See pages 1–3 for further explanation.

20. Lisa Nakamura, "Blaming, Shaming and the Feminization of Social Media," in *Feminist Surveillance Studies*, ed. Rachel E. Dubrofsky and Shoshana Amielle Magnet (Durham, NC: Duke University Press, 2015), 222.

21. Sherri Turkle, *Alone Together: Why We Expect More from Technology and Less from Each Other* (New York: Basic Books, 2012), 166.

22. Kerri Harvey, *Eden Online: Re-inventing Humanity in a Technological Universe* (New York: Hampton Press, 2000), 139.

23. Harvey, *Eden Online*, 140.

24. Phil Cooke, *Unique: Telling Your Story in the Age of Brands and Social Media* (Grand Rapids, MI: Baker Books, 2015), 19.

25. Cooke, *Unique*, 50.

26. Cooke, *Unique*, 50.

27. Haidt and Joseph, "The Moral Mind," 390.

28. Haidt and Joseph, "The Moral Mind," 390.

29. Marshall McLuhan, *Understanding Media: The Extensions of Man* (New York: McGraw-Hill, 1964). He later published *The Medium Is the Massage* using a text and graphic style for a wider lay audience. See Marshall McLuhan, Quentin Fiore, and Jerome Agel, *The Medium Is the Massage* (New York: Bantam Books, 1967).

30. Erving Goffmann, *The Presentation of Self in Everyday Life* (New York: Anchor Books, 1959).

31. Judith E. Rosenbaum, Benjamin K. Johnson, Peter A. Stepman, and Koos C. M. Nuijten, "'Looking the Part' and 'Staying True': Balancing Impression Management on Facebook," in *Social Networking and Impression Management: Self-Presentation in the Digital Age*, ed. Carolyn Cunningham (Lanham, MD: Lexington Books, 2013), 38.

32. Rosenbaum et al., "'Looking the Part,'" 54.

33. Bruce E. Drushel, "Virtual Closets: Strategic Identity Construction and Social Media," in *Social Networking and Impression Management: Self-Presentation in the Digital Age*, ed. Carolyn Cunningham (Lanham, MD: Lexington Books, 2013), 153–54.

34. Drushel, "Virtual Closets," 160.

35. Jeffrey A. Hall and Natalie Pennington, "What You Can Really Know about Someone from Their Facebook Profile (and Where You Should Look to Find Out)," in *Social Networking and Impression Management: Self-Presentation in the Digital Age*, ed. Carolyn Cunningham (Lanham, MD: Lexington Books, 2013), 248.

36. Donna Freitas, *The Happiness Effect: How Social Media Is Driving a Generation to Appear Perfect at Any Cost* (New York: Oxford University Press, 2017).

37. Freitas, *The Happiness Effect*, 48.

38. Corey Jay Liberman, "Branding as Social Discourse: Identity Construction Using Online Social and Professional Networking Sites," in *Social Networking and Impression Management: Self-Presentation in the Digital Age*, ed. Carolyn Cunningham (Lanham, MD: Lexington Books, 2013), 118.

39. danah boyd details the origin of the term on her site at http://www. zephoria.org/thoughts/archives/2013/12/08/coining-context-collapse.html.

40. Sara Green-Hamann and John C. Sherblom, "Developing a Transgender Identity in a Virtual Community," in *Social Networking and Impression Management: Self-Presentation in the Digital Age*, ed. Carolyn Cunningham (Lanham, MD: Lexington Books, 2013), 185.

41. See Jeffrey H. Kuznekoff, "Comparing Impression Management Strategies across Social Media Platforms," in *Social Networking and Impression Management: Self-Presentation in the Digital Age*, ed. Carolyn Cunningham (Lanham, MD: Lexington Books, 2013), 15–34.

42. There is much that can be said about issues of embodiment, gender, sexuality, and digital technology. Because there is not space in this text to address them all with the nuance and care needed, I have chosen not to touch on various issues, from selfies to sexting to online pornography use. However, it seems relevant here to at least note that there is a growing body of literature related to these issues and specific to the context of Christian faith traditions. Related to online pornography, I would encourage the reader to begin with Joyce Ann Mercer, "Pornography and the Abuse of Social Media," in *Professional Sexual Ethics*, ed. Patricia Jung and Darryl Stephens (Minneapolis, MN: Fortress Press, 2013), 193–203. I have published in online youth ministry blogs, curriculum, and videos covering issues of sexting and pornography that can be easily accessed via an online search.

43. Margaret A. Farley, *Just Love: A Framework for Christian Sexual Ethics* (New York: Continuum, 2006), 116–18.

44. Roberto Simanowski, *Data Love: The Seduction and Betrayal of Digital Technologies* (New York: Columbia University Press, 2016), 13.

45. Simanowski, *Data Love*, 13–14.

46. Simanowski, *Data Love*, 73.

47. John Cheney-Lippold, *We Are Data: Algorithms and the Making of Our Digital Selves* (New York: New York University Press, 2017), xiii.

48. Cheney-Lippold, *We Are Data*, 14.

49. Cheney-Lippold, *We Are Data*, 10.

50. Cheney-Lippold, *We Are Data*, 8.

51. Cheney-Lippold, *We Are Data*, 24.

52. Seth Stephens-Davidowitz, *Everybody Lies: Big Data, New Data, and What the Internet Can Tell Us about Who We Really Are* (New York: Harper-Collins, 2017), 4.

53. Stephens-Davidowitz, *Everybody Lies*, 112.

54. Stephens-Davidowitz, *Everybody Lies*, 14.

55. Stephens-Davidowitz, *Everybody Lies*, 123.

56. Stephens-Davidowitz, *Everybody Lies*, 123.

57. Jennifer Harvey, *Dear White Christians: For Those Still Longing for Racial Reconciliation* (Grand Rapids, MI: Eerdmans, 2014), 52.

58. Harvey, *Dear White Christians*, 56.

59. Harvey, *Dear White Christians*, 55.

60. Brad J. Kallenberg, *God and Gadgets: Following Jesus in a Technological Age* (Eugene, OR: Cascade Books, 2011), 117.

61. Kallenberg, *God and Gadgets*, 117.

62. Shane Hipps, *The Hidden Power of Electronic Culture: How Media Shapes Faith, the Gospel and Church* (Grand Rapids, MI: Zondervan, 2005), 23.

63. Freitas, *The Happiness Effect*, 58–61.

64. Freitas, *The Happiness Effect*, 61.

65. Freitas, *The Happiness Effect*, 61.

66. Freitas, *The Happiness Effect*, 110.

67. Freitas, *The Happiness Effect*, 110.

68. Quentin J. Schultze, *Habits of the High-Tech Heart: Living Virtuously in the Information Age* (Grand Rapids, MI: Baker Academic, 2002), 117 and 116, respectively.

69. Schultze, *Habits of the High-Tech*, 116.

70. Kallenberg, *God and Gadgets*, 46–81.

71. Gordon S. Mikoski, "Bringing the Body to the Table," *Theology Today* 67 (2010): 255–59.

72. Mikoski, "Body to the Table," 257.

73. Mikoski, "Body to the Table," 258.

74. Mikoski, "Body to the Table," 258.

75. Harvey, *Eden Online*, 135.

76. John Wall, *Ethics in Light of Childhood* (Washington, DC: Georgetown University Press, 2010), 169. The reader will recall this quote from the introductory section of the book related to new forms of ethics that need further development in a digital landscape.

77. See Nigel Hayes's Twitter account @NIGEL_HAYES, https://twitter.com/nigel_hayes/status/778730636529967104?lang=en.

78. Witnessing violence through media can contribute to vicarious traumatization and that social media and digital technology can deepen the trauma. While the exposure is not direct, it is often ongoing and ubiquitous given media saturation.

79. M. Shawn Copeland, *Enfleshing Freedom: Body, Race, and Being* (Minneapolis, MN: Fortress Press, 2010), 94.

80. Copeland, *Enfleshing Freedom: Body, Race, and Being*, 94.

## 3. MORAL FUNCTIONS BEYOND THE DELETE KEY

1.   Parts of this chapter appeared in the "The Internet Endureth Forever," *Sojourners*, March 2017, https://sojo.net/magazine/march-2017/internet-endureth-forever. Portions are adapted and reprinted with permission from *Sojourners* (800-714-7474, https://sojo.net).

2.   Lisa Nakamura, "Blaming, Shaming, and the Feminization of Social Media," in *Feminist Surveillance Studies*, ed. Rachel E. Dubrofsky and Shoshana Amielle Magnet (Durham, NC: Duke University Press, 2015), 224.

3.   Rosemary Radford Ruether, "Feminist Metanoia and Soul-Making," in *Women's Spirituality: Women's Lives*, ed. Judith Ochshorn and Ellen Cole (Philadelphia: Haworth Press, 1995), 38.

4.   Ruether, "Feminist Metanoia," 33.

5.   Ruether, "Feminist Metanoia," 34.

6.   See Emilie M. Townes, *Womanist Ethics and the Cultural Production of Evil* (New York: Palgrave Macmillan, 2006), for a fuller discussion of historical and cultural production of evil as well as how individual actions comprise systemic social sin.

7.   Ruether, "Feminist Metanoia," 34.

8.   See Verity Jones, "Living Theologically in a Networked World," *Reflections*, accessed August 5, 2016, http://reflections.yale.edu/article/ibelieve-facing-new-media-explosion/living-theologically-networked-world.

9.   Johanna Blakley, "Media in Our Image," *Women's Studies Quarterly* 40, nos. 1 & 2 (Spring/Summer 2012): 343.

10.   See John Dyer, *From the Garden to the City: The Redeeming and Corrupting Power of Technology* (Grand Rapids, MI: Kregel Publications, 2011), 24–25; and Astra Taylor, *The People's Platform: Taking Back Power and Culture in the Digital Age* (New York: Metropolitan Books, 2014), 76–78.

11.   See danah boyd, "Social Network Sites and Networked Publics: Affordances, Dynamics, and Implications," in *A Networked Self: Identity, Community, and Culture on Social Network Sites*, ed. Zizi Papacharissi (New York: Routledge, 2010), 39–58; and danah boyd, *It's Complicated: The Social Lives of Networked Teens* (New Haven, CT: Yale University Press, 2014), 10–11.

12.   Mary Madden and Lee Rainie, "Americans' Attitudes about Privacy, Security and Surveillance," Pew Research Center, May 20, 2015, accessed November 2, 2016, http://www.pewinternet.org/2015/05/20/americans-attitudes-about-privacy-security-and-surveillance.

13.   Juan Enriquez, "Your Online Life, Permanent as a Tattoo," TED Talk, February 2013, accessed September 21, 2016, https://www.ted.com/talks/juan_enriquez_how_to_think_about_digital_tattoos.

14. Viktor Mayer-Schönberger and Kenneth Cukier, *Big Data: A Revolution That Will Transform How We Live, Work, and Think* (New York: Houghton Mifflin Harcourt, 2013), 5.

15. Alice E. Marwick, "The Public Domain: Surveillance in Everyday Life," *Surveillance and Society* 9, no. 4 (2012): 382.

16. Marwick, "The Public Domain," 386.

17. See Rachel E. Dubrofsky and Shoshana Amielle Magnet, eds., *Feminist Surveillance Studies* (Durham, NC: Duke University Press, 2015). This collection of essays uses "feminist theory to expose the ways in which surveillance practices and technologies are tied to systemic forms of discrimination that serve to normalize whiteness, able-bodiedness, capitalism and heterosexuality" (from the book description).

18. See "Anita Sarkeesian," *Wikipedia*, accessed October 19, 2016, https://en.wikipedia.org/wiki/Anita_Sarkeesian.

19. Katie Rogers, "Leslie Jones, Star of *Ghostbusters*, Becomes a Target of Online Trolls," *New York Times*, July 19, 2016, accessed October 19, 2016, https://www.nytimes.com/2016/07/20/movies/leslie-jones-star-of-ghostbusters-becomes-a-target-of-online-trolls.html.

20. See the Stalking Resource Center for the most up-to-date research for prosecutors, law enforcement officials, and practitioners, http://victimsofcrime.org/our-programs/stalking-resource-center.

21. Barney Calman, "Fibbing on Facebook Can Trick Your Memory: People Start Believing Their Own Social Media Exagerations," *Daily Mail*, December 24, 2014, accessed October 19, 2016, http://www.dailymail.co.uk/health/article-2888454/Youngsters-airbrushing-reality-social-media-make-lives-interesting-suffer-paranoia-sadness-shame-fail-live-online-image.html.

22. Marwick, "The Public Domain," 390.

23. Quentin J. Schultze, *Habits of the High-Tech Heart: Living Virtuously in the Information Age* (Grand Rapids, MI: Baker Academic, 2004). In chapter 1, "Discerning Our Informationalism," Schultze raises questions about how informationalism shapes users to prefer the present, act as impersonal observers, and find reward in measurement. These preferences lack intimacy, consequence, and meaning for Schultze. While he is concerned about issues of informational literacy, he is also raising meta-level questions about how information flow affects virtue development.

24. See European Commission, "Factsheet on Right to Be Forgotten Ruling," accessed July 30, 2018, https://www.inforights.im/media/1186/cl_eu_commission_factsheet_right_to_be_forgotten.pdf.

25. See Katie Halper, "A Brief History of People Getting Fired for Social Media Stupidity," *Rolling Stone*, July 13, 2015, accessed August 1, 2016, http://

www.rollingstone.com/culture/lists/a-brief-history-of-people-getting-fired-for-social-media-stupidity-20150713.

26. Kathleen Hennessey, "Rep. Anthony Weiner Makes Resignation Official," *Los Angeles Times*, June 20, 2011, accessed August 3, 2016, http://articles.latimes.com/2011/jun/20/news/la-pn-weiner-resignation-20110620.

27. Tom McCarty, "New York Mayoral Candidate Anthony Weiner Says Explicit Photo Is of Him," *Guardian*, July 23, 2013, accessed August 3, 2016, https://www.theguardian.com/world/2013/jul/23/anthony-weiner-new-york-explicit-photographs.

28. John Breech, "Ravens RB Ray Rice Indicted on One Count of Aggravated Assault," CBS Sports.com, March 27, 2014, accessed August 3, 2016, https://www.cbssports.com/nfl/news/ravens-rb-ray-rice-indicted-on-one-count-of-aggravated-assault.

29. "Ray Rice Elevator Knockout Fiancée Takes Crushing Punch," TMZsports.com, September 8, 2014, accessed August 3, 2016, http://www.tmz.com/2014/09/08/ray-rice-elevator-knockout-fiancee-takes-crushing-punch-video.

30. Louis Bien, "A Complete Timeline of the Ray Rice Assault Case," SBNation.com, November 28, 2014, accessed August 3, 2016, https://www.sbnation.com/nfl/2014/5/23/5744964/ray-rice-arrest-assault-statement-apology-ravens.

31. Meg Leta Jones, *Ctrl+Z: The Right to Be Forgotten* (New York: New York University Press, 2016).

32. Viktor Mayer-Schönberger, *Delete: The Virtue of Forgetting in the Digital Age* (Princeton, NJ: Princeton University Press, 2011), 12.

33. Ellie Kaufman and Jean Casarez, "Anthony Weiner Gets 21 Months in Prison in Sexting Case," CNN, September 25, 2017, accessed July 28, 2018, https://www.cnn.com/2017/09/25/politics/anthony-weiner-sentencing/index.html.

34. Ruether, "Feminist Metanoia," 42.

35. See Jeffrie G. Murphy, *Getting Even: Forgiveness and Its Limits* (New York: Oxford University Press, 2003).

36. See Murphy, *Getting Even*, for further discussion of forgiveness as distinct from mercy or reconciliation. See also Beverly Wildung Harrison, "The Power of Anger in the Work of Love," in *Making the Connections: Essays in Feminist Social Ethics*, ed. Carol S. Robb (Boston: Beacon Press, 1985), 3–21, for how the emotion of anger is evidence of something morally amiss related to both personal experience and systemic sin.

37. Ruether, "Feminist Metanoia," 35.

38. See Privacy International for more information, https://www.privacyinternational.org.

39. Nakamura, "Feminization of Social Media," 226–27.


# 4. CREATION CONNECTIVITY

1. I realize that the terms *ecological* and *environmental* have specific contextual uses and both often refer to living organisms, with environmental issues specifically involving humans. Throughout the chapter, I will use these terms in relation to digital technology to push the definitions in ways that suggest digital technologies can have an impact similar to organisms as newer technologies take on vibrational and connective relationships that are part of biodiversity in a digitized world.

2. Rosemary Radford Ruether, *Gaia and God: An Ecofeminist Theology of Earth Healing* (San Francisco: HarperCollins, 1992), 86.

3. See University of Michigan's Center for Sustainable Systems factsheets and research, especially the US environmental footprint factsheet, at http://css.umich.edu/factsheets/us-environmental-footprint-factsheet, and discussion of research from the University of California–Berkeley in *Science Daily* on "Rich Nations' Environmental Footprints Tread Heavily on Poor Countries," January 23, 2008, accessed on September 3, 2017, https://www.sciencedaily.com/releases/2008/01/080121181408.htm. For discussion of the theological and ethical issues relevant to social disparities of the ecological crisis, see Christiana Z. Peppard, Julia Watts Belser, Erin Lothes Biviano, and James B. Martin-Schramm, "What Powers Us? A Comparative Religious Ethics of Energy Sources, Power, and Privilege," *Journal of the Society of Christian Ethics* 36, no. 1 (2016).

4. Ruether, *Gaia and God*, 259. See pages 26–31 for a full discussion of Ruether's response to immanence and transcendence.

5. Sallie McFague, *The Body of God: An Ecological Theology* (Minneapolis, MN: Fortress Press, 1993), 20.

6. Ivone Gebara, *Longing for Running Water: Ecofeminism and Liberation*, trans. David Molineaux (Minneapolis, MN: Fortress Press, 1999), 97.

7. Melanie L. Harris, *Ecowomanism: African American Women and Earth-Honoring Faiths* (Maryknoll, NY: Orbis, 2017), 1.

8. Harris, *Ecowomanism*, 22.

9. Pope Francis, "Encyclical Letter Laudato Si' of the Holy Father Francis on Care for Our Common Home" (The Holy See: Vatican Press, 2015), #48.

10. Pope Francis, "Laudato Si'," #49.

11. Pope Francis, "Laudato Si'," #49.

12. Willis Jenkins, *The Future of Ethics: Sustainability, Social Justice, and Religious Creativity* (Washington, DC: Georgetown University Press, 2013), 3.

13. Jenkins, *Future of Ethics*, 3.

14. Jenkins, *Future of Ethics*, 232.

15. Gili S. Drori, *Global E-Litism: Digital Technology, Social Inequality, and Transnationality* (New York: Macmillan, 2005), 8.

16. Pope Francis, "Laudato Si'," #20.

17. Pope Francis, "Laudato Si'," #107.

18. Jennifer Gabrys, "Powering the Digital: From Energy Ecologies to Electronic Environmentalism," in *Media and the Ecological Crisis*, ed. Richard Maxwell, Jon Raundalen, and Nina Lager Vestberg (New York and London: Routledge, 2014), 3–18.

19. Jussi Parikka, *Anthrobscene* (Minneapolis, MN: University of Minnesota Press, 2014), 5–6.

20. See "A World of Minerals in Your Mobile Device," US Geological Survey, September 2016, accessed August 2017, https://pubs.usgs.gov/gip/0167/gip167.pdf.

21. Parikka, *Anthrobscene*, 55.

22. Parikka, *Anthrobscene*, 56.

23. Muhirgirwa Rusembuka Ferdinand, SJ, "Theological Perspectives on Governance in the Mining Sector in the Democratic Republic of Congo," in *Just Sustainability: Technology, Ecology, and Resource Extraction*, ed. Christiana Peppard and Andrea Vincini, SJ (Maryknoll, NY: Orbis Books, 2015), Loc 895 of 6545.

24. T. V. Reed, *Digitized Lives: Culture, Power, and Social Change in the Internet Era* (New York: Routledge, 2014), 46.

25. Peter Knox, SJ, "Sustainable Mining in South Africa: A Concept in Search of a Theory," in *Just Sustainability: Technology, Ecology, and Resource Extraction*, ed. Christiana Peppard and Andrea Vincini, SJ (Maryknoll, NY: Orbis Books, 2015), Loc 2779 of 6545.

26. Ferdinand, "Theological Perspectives on Governance," Loc 815 of 6545. For updates and information related to conflict minerals, see "Progress and Challenges on Conflict Minerals," !enough, https://enoughproject.org/special-topics/progress-and-challenges-conflict-minerals-facts-dodd-frank-1502.

27. Reed, *Digitized Lives*, 46.

28. Pope Francis, "Laudato Si'," #21.

29. Pope Francis, "Laudato Si'," #22.

30. Parikka, *Anthrobscene*, 16.

31. Parikka, *Anthrobscene*, 36.

32. See Parikka, *Anthrobscene*, 43, quoting Martin Heidegger, *The Question Concerning Technology and Other Essays*, trans. William Lovitt (New York: Garland Publishing, 1977), 16. Also, see Brian Brock, *Christian Ethics in a Technological Age* (Grand Rapids, MI: Eerdmans, 2010), for an extended

investigation and critique of Heidegger's theory as it relates to ethical claims and Christian theology.

33. Gabrys, "Powering the Digital," 3–18.

34. Gabrys, "Powering the Digital," 5.

35. Kevin Slavin, "How Algorithms Shape Our World," TEDGlobal 2011, accessed January 20, 2017, https://www.ted.com/talks/kevin_slavin_how_ algorithms_shape_our_world. In a connection to other chapters in this book, Slavin remarks that algorithms will be the third evolutionary force. Algorithms, which run software programs—the programs that return a browser search, for example—require servers to process data more quickly. Slavin says, "We'll actually part the water to pull money out of the air because it's a bright future if you're an algorithm" (13:11–14:00). We are unaware of the "seismic, terrestrial effects of the math we are making," comments Slavin (14:02–14:50).

36. Athima Chansanchai, "Microsoft Research Project Puts Cloud in Ocean for the First Time," Microsoft News, February 1, 2016, accessed August 31, 2017, https://news.microsoft.com/features/microsoft-research-project-puts-cloud-in-ocean-for-the-first-time. See also Bidness Etc., "Microsoft's New Data Centers Will Hum Deep under the Sea," YouTube, February 1, 2016, https://www.youtube.com/watch?v=z4NlG4hKAvk.

37. Brian Anderson, "Data Centers Are Really Freaking Loud," Motherboard, May 2, 2014, accessed August 31, 2017, https://motherboard.vice.com/en_us/article/z4m4qy/data-centers-are-really-freaking-loud.

38. Michael Kassner, "Data Centers May Be Hazardous to Your Hearing," Tech Republic, February 8, 2014, accessed August 31, 2017, http://www.techrepublic.com/article/data-centers-may-be-hazardous-to-your-hearing.

39. See Autodestructo on Soundcloud as an example. https://soundcloud.com/autodestructo/dc1.

40. Samuel Arbesman, *Overcomplicated: Technology at the Limits of Comprehension* (New York: Current, 2016), 3–4.

41. Viviane Minikongo Mundele, "Ecology, Moral Theology, and Spirituality: A Perspective from the Democratic Republic of Congo," in *Just Sustainability: Technology, Ecology, and Resource Extraction*, ed. Christiana Peppard and Andrea Vincini, SJ (Maryknoll, NY: Orbis Books, 2015), Loc 732 of 6545.

42. Gabrys, "Powering the Digital," 5.

43. Jenkins, *Future of Ethics*, 241.

44. Jenkins, *Future of Ethics*, 242.

45. Jenkins, *Future of Ethics*, 243.

46. Jenkins, *Future of Ethics*, 258.

47. See the Citizen Sense Project's home page for more descriptions of its work and project stories, accessed January 20, 2017, http://citizensense.net.

Direct quotes from the "About Citizen Sense" page, accessed August 31, 2017, http://citizensense.net/about.

48. See Citizen Sense Project's "About" page, accessed August 31, 2017, http://citizensense.net/about.

49. Jennifer Gabrys, *Program Earth: Environmental Sensing Technology and the Making of a Computational Planet* (Minneapolis, MN: University of Minnesota Press, 2016), 67.

50. Gabrys, *Program Earth*, 269.

51. Gabrys, *Program Earth*, 268.

52. Jenkins, *Future of Ethics*, 261.

53. John Sniegocki, "The Political Economy of Sustainability," in *Just Sustainability: Technology, Ecology, and Resource Extraction*, ed. Christiana Peppard and Andrea Vincini, SJ (Maryknoll, NY: Orbis Books, 2015), loc. 1344 of 6545.

54. Richard Maxwell and Toby Miller, "The Waste of Art and the Art of Waste: The Problems with 'You' and Your Media," *Psychology Today*, October 6, 2013, accessed September 2, 2017, https://www.psychologytoday.com/blog/greening-the-media/201310/the-waste-art-and-the-art-waste.

55. Reed, *Digitized Lives*, 51.

56. Reed, *Digitized Lives*, 46.

57. Reed, *Digitized Lives*, 48.

58. Maxwell and Miller, "The Waste of Art and the Art of Waste."

59. See Leonard's website, which is dedicated to the Story of Stuff Project, accessed March 20, 2017, http://storyofstuff.org/annie.

60. See the web page for Phone Story app, accessed September 3, 2017, http://www.phonestory.org.

61. See Jordan's website, accessed September 3, 2017, http://www.artworksforchange.org/portfolio/chris-jordan.

62. See Jordan's portfolio, accessed September 3, 2017, http://www.artworksforchange.org/portfolio/chris-jordan.

63. For additional resources related to the art of e-waste, accessed September 3, 2017, https://www.betterworldsolutions.eu/e-waste-as-art/, https://www.trendhunter.com/protrends/e-artisans, and http://www.bbc.com/future/story/20140218-why-your-old-tech-holds-treasure.

64. Pope Francis, "Laudato Si'," #113.

65. Harris, *Ecowomanism*, 149.

66. Jenkins, *Future of Ethics*, 233.

67. McFague, *The Body of God*, 133.

# 5. ETHICAL HACKING AND
# HACKING ETHICS

1. Isaiah 2:4, translation New Revised Standard Version.

2. Patricia K. Tull, "Isaiah," in *Women's Bible Commentary*, 2nd ed., ed. Carol Ann Newsom, Sharon H. Ringe, and Jacqueline E. Lapsley (Louisville, KY: Westminster John Knox, 2012), 258. Tull's feminist commentary on Isaiah deals with the problematic female imagery employed by Isaiah, which I do not deal with in this chapter.

3. See RAWtools website, "About Us," accessed April 20, 2017, http://rawtools.org/about-us.

4. See RAWtools website, "About Us."

5. Tull, "Isaiah," 255.

6. Barry M. Leiner, Robert E. Kahn, Jon Postel, Vinton G. Cerf, Leonard Kleinrock, David D. Clark, Daniel C. Lynch, Stephen Wolff, and Larry G. Roberts, "A Brief History of the Internet," *Computer Communication Review* 39, no. 5 (October 2009): 22. At the time of Licklider's leadership, DARPA was called ARPA. Given the reader's context, I have chosen to use the more common name DARPA as the one that has continued to be used more often.

7. Leiner et al., "A Brief History of the Internet," 29.

8. Edward Le Roy Long Jr., "The Moral Assessment of Computer Technology," in *The Public Vocation of Christian Ethics*, ed. Beverly W. Harrison, Robert L. Stivers, and Ronald H. Stone (New York: Pilgrim Press, 1986), 244.

9. Long, "The Moral Assessment of Computer Technology," 242.

10. Andrew V. Edwards, *Digital Is Destroying Everything: What the Tech Giants Won't Tell You about How Robots, Big Data, and Algorithms Are Radically Remaking Your Future* (Lanham, MD: Rowman & Littlefield, 2015), 201.

11. Rodney H. Jones and Christoph A. Hafner, *Understanding Digital Literacies: A Practical Introduction* (New York: Routledge, 2012), 5–9.

12. Jones and Hafner, *Understanding Digital Literacies*, 111.

13. Jones and Hafner, *Understanding Digital Literacies*, 7.

14. Jones and Hafner, *Understanding Digital Literacies*, 190. The alteration of the quote from the use of "mastering" to "cultivating" is mine. It is an attempt to "hack" Jones and Hafner in the spirit of their own ethical arguments. "Master" carries a connotation of dominance and a material connection to master–slave relationships that exploited ancestors of Black and Brown people in the United States and those various forms of slavery that continue throughout the world today. Additionally, the use of the word suggests a human practice with regard to literacy that I view as unrealistic and purposeless. Mastery suggests there is no need for ongoing growth.

15. Jones and Hafner, *Understanding Digital Literacies*, 191.

16. Jones and Hafner, *Understanding Digital Literacies*, 191. The alteration of the quote from the use of "master" to "cultivate" is mine. See note 14 for further explanation.

17. Brett Lunceford, "Programs of People? Participation and the Ethics of Hacktivism," in *Controversies in Digital Ethics*, ed. Amber Davisson and Paul Booth (New York: Bloomsbury, 2016), 77.

18. Lunceford, "Programs of People?," 82.

19. Lunceford, "Programs of People?," 88.

20. For more information on the topic of net neutrality, see Moz://a Podcast Episode 2, "The Neutral Zone: The Future of Net Neutrality," accessed September 28, 2017, https://irlpodcast.org/episode2.

21. Some portions of the following section related to Twitter were previously published as the "In the News: Twitter Activism" 2014 adult study guide for TheThoughtfulChristian.com, https://www.thethoughtfulchristian.com/Products/Default.aspx?bookid=TC0584. Portions are adapted and reprinted with permission from Westminster John Knox Press and TheThoughtfulChristian.com.

22. Feminista Jones, "Is Twitter the Underground Railroad of Activism?," *Salon*, July 17, 2007, accessed June 4, 2014, http://www.salon.com/2013/07/17/how_twitter_fuels_black_activism.

23. Maeve Duggan and Joanna Brenner, "The Demographics of Media Users—2012," Pew Research Center, February 14, 2013, accessed May 28, 2014, http://www.pewinternet.org/files/old-media//Files/Reports/2013/PIP_SocialMediaUsers.pdf.

24. @Prisonculture and Andrea Smith, "Interlopers on Social Media: Feminism, Women of Color and Oppression," Prison Culture, January 30, 2014, accessed June 3, 2014, http://www.usprisonculture.com/blog/2014/01/30/interlopers-on-social-media-feminism-women-of-color-and-oppression.

25. Andrea Grimes, "Hashtag Activism and the Lie of 'Solidarity,'" Rewire.News, May 28, 2014, accessed May 28, 2014, https://rewire.news/article/2014/05/28/hashtag-activism-lie-solidarity. For further discussion of activism and feminism across online communities, see Kate Ott, "Social Media and Feminist Values: Aligned or Maligned?," *Frontiers: A Journal of Women's Studies* 39, no. 11 (2018): 93–111.

26. Anthony McCosker, Sonja Vivienne, and Amelia Johns, eds. *Negotiating Digital Citizenship: Control, Contest and Culture* (London: Rowman & Littlefield International, 2016), 1.

27. McCosker, Vivienne, and Johns, *Negotiating Digital Citizenship*, 7.

28. McCosker, Vivienne, and Johns, *Negotiating Digital Citizenship*, 10.

29. John Wall, *Ethics in Light of Childhood* (Washington, DC: Georgetown University Press, 2010), 169.

30.  Matthew Pittman and Tom Bivins, "Just War Craft: Virtue Ethics and DotA," in *Controversies in Digital Ethics*, ed. Amber Davisson and Paul Booth (New York: Bloomsbury, 2016), 83.

31.  Henry Jenkins, Mizuko Ito, and danah boyd, *Participatory Culture in a Networked Era: A Conversation on Youth, Learning, Commerce, and Politics* (Malden, MA: Polity Press, 2016), 181. For a fuller definition of participatory culture, see Henry Jenkins's blog, accessed October 1, 2017, http://henryjenkins.org/blog/2006/10/confronting_the_challenges_of.html. He defines participatory culture as:

1. With relatively low barriers to artistic expression and civic engagement

2. With strong support for creating and sharing one's creations with others

3. With some type of informal mentorship whereby what is known by the most experienced is passed along to novices

4. Where members believe that their contributions matter

5. Where members feel some degree of social connection with one another (at the least they care what other people think about what they have created).

32.  Jenkins, Ito, and boyd, *Participatory Culture*, 182.

33.  See Jenkins, Ito, and boyd, *Participatory Culture*, for examples related to youth. For a wide variety of participatory communities engaged in platform cooperativism, see *Ours to Hack and to Own: The Rise of Platform Cooperativism, a New Vision for the Future of Work and a Fairer Internet*, ed. Terbor Scholz and Nathan Schneider (New York: OR Books, 2016).

34.  The information provided on the ESLA comes from its website and an interview with Rev. Megan Hollaway, chaplain and head of the Religion and Ethics Department. The interview took place via phone on September 8, 2017. No direct quotes are used from that phone call. It was informational in nature, and Rev. Hollaway is a public figure as both a religious leader and current acting leader of the institution while the founder and head of the school, Rev. Maryetta Anschutz, is on sabbatical. For more information on the school, see its website, http://es-la.com/. Rev. Anschutz is a former Yale Divinity School classmate and current colleague of mine. This is how I discovered the school and followed its growth.

35.  See https://youtu.be/RP14C6BG1Tk for the YouTube video, which is also featured on the website, http://es-la.com/#and-a-community-grows. The blog is no longer available to public viewers.

36. See Technology and Pedagogy, http://es-la.com/#makes-a-world-of-difference.

37. See Our Episcopal Identity, http://es-la.com/#to-look-toward-the-future.

38. Jones and Hafner, *Understanding Digital Literacies*, 191. The alteration of the quote from the use of "master" to "cultivate" is mine. See note 14 for further explanation.

39. Amy C. Merrill Willis, "Habakkuk," in *Women's Bible Commentary*, ed. Carol A. Newsom, Sharon H. Ringe, and Jacqueline E. Lapsley (Louisville, KY: Westminster John Knox, 2012), 336.

40. Judith E. Sanderson, "Habakkuk," in *Women's Bible Commentary*, ed. Carol A. Newsom and Sharon H. Ringe (Louisville, KY: Westminster John Knox, 1998), 238.

41. Theodore Hiebert, "The Book of Habakkuk," in *The New Interpreter's Bible: Old Testament Survey*, ed. Beverly Roberts Gaventa and David Petersen (Nashville: Abingdon Press, 2005), 442.

42. Hiebert, "The Book of Habakkuk," 442.

# SELECTED BIBLIOGRAPHY

Arbesman, Samuel. *Overcomplicated: Technology at the Limits of Comprehension*. New York: Current, 2016.

boyd, danah. *It's Complicated: The Social Lives of Networked Teens*. New Haven, CT: Yale University Press, 2014.

Brock, Brian. *Christian Ethics in a Technological Age*. Grand Rapids, MI: Eerdmans, 2010.

Burn, Andrew. *Making New Media: Creative Production and Digital Literacies*. New York: Peter Lang Publishing, 2009.

Carruthers, Peter, Stephen Laurence, and Stephen Stich, eds. *The Innate Mind, Volume 3: Foundations and the Future*. Oxford: Oxford University Press, 2008.

Cessario, Romanus, OP. *The Moral Virtues and Theological Ethics*. Notre Dame, IN: Notre Dame Press, 2009.

Cheney-Lippold, John. *We Are Data: Algorithms and the Making of Our Digital Selves*. New York: New York University Press, 2017.

Cheong, Pauline Hope, Peter Fisher-Nielsen, Stefan Gelfgren, and Charles Ess, eds. *Digital Media, Social Media and Culture: Perspectives, Practices and Futures*. New York: Peter Lang Publishing, 2012.

Cooke, Phil. *Unique: Telling Your Story in the Age of Brands and Social Media*. Grand Rapids, MI: Baker Books, 2015.

Copeland, M. Shawn. *Enfleshing Freedom: Body, Race, and Being*. Minneapolis, MN: Fortress Press, 2010.

Cunningham, Carolyn, ed. *Social Networking and Impression Management: Self-Presentation in the Digital Age*. Lanham, MD: Lexington Books, 2013.

Drori, Gili S. *Global E-Litism: Digital Technology, Social Inequality, and Transnationality*. New York: Macmillan, 2005.

Dubrofsky, Rachel E., and Shoshana Amielle Magnet. *Feminist Surveillance Studies*. Durham, NC: Duke University Press, 2015.

Dyer, John. *From the Garden to the City: The Redeeming and Corrupting Power of Technology*. Grand Rapids, MI: Kregel Publications, 2011.

Edwards, Andrew V. *Digital Is Destroying Everything: What the Tech Giants Won't Tell You about How Robots, Big Data, and Algorithms Are Radically Remaking Your Future*. Lanham, MD: Rowman & Littlefield, 2015.

Eubanks, Virginia. *Automating Inequality: How High-Tech Tools Profile, Police, and Punish the Poor*. New York: St. Martin's Press, 2017.

Freire, Paulo. *Pedagogy of the Oppressed*. 30th anniversary ed. Translated by Myra Bergman Ramos. New York: Continuum, 2005.

Freitas, Donna. *The Happiness Effect: How Social Media Is Driving a Generation to Appear Perfect at Any Cost*. New York: Oxford University Press, 2017.

Friesen, Dwight J. *Thy Kingdom Connected: What the Church Can Learn from Facebook, the Internet, and Other Networks*. Grand Rapids, MI: Baker Books, 2009.

Gaventa, Beverly Roberts, and David Petersen, eds. *The New Interpreter's Bible: Old Testament Survey*. Nashville, TN: Abingdon Press, 2010.

Gebara, Ivone. *Longing for Running Water: Ecofeminism and Liberation*. Translated by David Molineaux. Minneapolis, MN: Fortress Press, 1999.

Gula, Richard M. *Just Ministry: Professional Ethics for Pastoral Ministers*. Mahwah, NJ: Paulist Press, 2010.

Harris, Melanie L. *Ecowomanism: African American Women and Earth-Honoring Faiths*. Maryknoll, NY: Orbis, 2017.

Hartley, John. *The Uses of Digital Literacy*. New Brunswick, NJ: Transaction Publishers, 2012.

Harvey, Jennifer. *Dear White Christians: For Those Still Longing for Racial Reconciliation*. Grand Rapids, MI: Eerdmans, 2014.

Harvey, Kerri. *Eden Online: Re-inventing Humanity in a Technological Universe*. New York: Hampton Press, 2000.

Hipps, Shane. *The Hidden Power of Electronic Culture: How Media Shapes Faith, the Gospel and Church*. Grand Rapids, MI: Zondervan, 2005.

Hunsinger, Jeremy, and Theresa Senft, eds. *The Social Media Handbook*. New York: Routledge, 2014.

"iBelieve: Facing the New Media Explosion." *Reflections* (Fall 2011). https://reflections.yale.edu/archives-list/2011.

Jenkins, Henry, Mizuko Ito, and danah boyd. *Participatory Culture in a Networked Era: A Conversation on Youth, Learning, Commerce, and Politics*. Malden, MA: Polity Press, 2016.

Jenkins, Willis. *The Future of Ethics: Sustainability, Social Justice, and Religious Creativity*. Washington, DC: Georgetown University Press, 2013.

Jones, Rodney H., and Christoph A. Hafner. *Understanding Digital Literacies: A Practical Introduction*. New York: Routledge, 2012.

Kallenberg, Brad J. *God and Gadgets: Following Jesus in a Technological Age*. Eugene, OR: Cascade Books, 2011.

Keenan, F. James, SJ. *Virtues for Ordinary Christians*. Franklin, WI: Sheed & Ward, 1996.

Levison, John R., and Priscilla Pope-Levison, eds. *Return to Babel: Global Perspectives on the Bible*. Louisville, KY: Westminster John Knox, 1999.

Levmore, Saul, and Martha Nussbaum, eds. *The Offensive Internet: Speech, Privacy, and Reputation*. Boston: Harvard University Press, 2010.

Long, Edward LeRoy, Jr. "The Moral Assessment of Computer Technology." In *The Public Vocation of Christian Ethics*, edited by Beverly W. Harrison, Robert L. Stivers, and Ronald H. Stone, 241–55. New York: Pilgrim Press, 1986.

Lule, Jack. *Globalization & Media: Global Village of Babel*. Lanham, MD: Rowman & Littlefield, 2012.

Magnet, Shoshana Amielle. *When Biometrics Fail*. Chapel Hill: Duke University Press, 2011.

Marwick, Alice E. "The Public Domain: Surveillance in Everyday Life." *Surveillance and Society* 9, no. 4 (2012): 378–93.

Maxwell, Richard, Jon Raundalen, and Nina Lager Vestberg, eds. *Media and the Ecological Crisis*. New York and London: Routledge, 2014.

Mayer-Schönberger, Viktor. *Delete: The Virtue of Forgetting in the Digital Age*. Princeton, NJ: Princeton University Press, 2011.

Mayer-Schönberger, Viktor, and Kenneth Cukier. *Big Data: A Revolution That Will Transform How We Live, Work, and Think*. New York: Houghton Mifflin Harcourt, 2013.

McCosker, Anthony, Sonja Vivienne, and Amelia Johns, eds. *Negotiating Digital Citizenship: Control, Contest and Culture*. London: Rowman & Littlefield International, 2016.

McFague, Sallie. *The Body of God: An Ecological Theology*. Minneapolis, MN: Fortress Press, 1993.

Murphy, Jeffrie G. *Getting Even: Forgiveness and Its Limits*. New York: Oxford University Press, 2003.

Newsom, Carol Ann, Sharon H. Ringe, and Jacqueline E. Lapsley, eds. *Women's Bible Commentary*. 2nd ed. Louisville, KY: Westminster John Knox, 2012.

Noble, Safiya Umoja. *Algorithms of Oppression: How Search Engines Reinforce Racism*. New York: New York University Press, 2018.

Noble, Safiya Umoja, and Brendesha M. Tynes, eds. *The Intersectional Internet: Race, Sex, Class, and Culture Online*. New York: Peter Lang Publishing, 2016.

Ott, Kate. "Children as An/other Subject: Redefining Moral Agency in a Postcolonial Context." *Journal of Childhood and Religion* 5, no. 2 (May 2014): 1–23.

———. "Social Media and Feminist Values: Aligned or Maligned?" *Frontiers: A Journal of Women's Studies* 39, no. 11 (2018): 93–111.

Papacharissi, Zizi, ed. *A Networked Self: Identity, Community, and Culture on Social Network Sites*. New York: Routledge, 2010.

Parikka, Jussi. *Anthrobscene*. Minneapolis, MN: University of Minnesota Press, 2014.

Pariser, Eli. *The Filter Bubble: What the Internet Is Hiding from You*. New York: Penguin, 2011.

Peppard, Christiana, and Andrea Vincini, SJ, eds. *Just Sustainability: Technology, Ecology, and Resource Extraction*. Maryknoll, NY: Orbis Books, 2015.

Pope Francis. "Encyclical Letter Laudato Si' of the Holy Father Francis on Care for Our Common Home." The Holy See: Vatican Press, 2015.

Reed, T. V. *Digitized Lives: Culture, Power, and Social Change in the Internet Era*. New York: Routledge, 2014.

Ruether, Rosemary Radford. "Feminist Metanoia and Soul-Making." In *Women's Spirituality: Women's Lives* edited by Judith Ochshorn and Ellen Cole, 33–44. Philadelphia: Haworth Press, 1995.

———. *Gaia and God: An Ecofeminist Theology of Earth Healing*. San Francisco: HarperCollins, 1992.

Scholz, Terbor, and Nathan Schneider. *Ours to Hack and to Own: The Rise of Platform Cooperativism, a New Vision for the Future of Work and a Fairer Internet*. New York: OR Books, 2016.

Schultze, Quentin J. *Habits of the High-Tech Heart: Living Virtuously in the Information Age*. Grand Rapids, MI: Baker Academic, 2004.

Simanowski, Roberto. *Data Love: The Seduction and Betrayal of Digital Technologies*. New York: Columbia University Press, 2016.

Stephens-Davidowitz, Seth. *Everybody Lies: Big Data, New Data, and What the Internet Can Tell Us about Who We Really Are*. New York: HarperCollins, 2017.

Taylor, Astra. *The People's Platform: Taking Back Power and Culture in the Digital Age*. New York: Metropolitan Books, 2014.

Townes, Emilie M. *Womanist Ethics and the Cultural Production of Evil*. New York: Palgrave Macmillan, 2006.

Traina, Cristina L. H. *Erotic Attunement: Parenthood and the Ethics of Sensuality between Unequals*. Chicago: Chicago University Press, 2001.

Turkle, Sherri. *Alone Together: Why We Expect More from Technology and Less from Each Other*. New York: Basic Books, 2012.

Wall, John. *Ethics in Light of Childhood*. Washington, DC: Georgetown University Press, 2010.

Wogaman, J. Philip. *Christian Ethics: A Historical Introduction*. Louisville, KY: Westminster John Knox, 1993.

## WEB CITATIONS

Algorithmic Fairness, http://fairness.haverford.edu/index.html

Algorithmic Justice League, https://www.ajlunited.org

Artworks for Change, http://www.artworksforchange.org

Citizen Sense Project, http://citizensense.net

Pew Research Center: Internet and Technology, http://www.pewinternet.org

Phone Story App, http://www.phonestory.org

Privacy International, https://www.privacyinternational.org

RAWtools, https://rawtools.org

Social Media Collective (SMC), Microsoft Research Labs, "Critical Algorithm Studies: A Reading List," https://socialmediacollective.org/reading-lists/critical-algorithm-studies

Story of Stuff Project, http://storyofstuff.org/annie

# INDEX